Stem Cells in Clinical Applications

Series Editor

Phuc Van Pham
VNUHCM-US Stem Cell Institute, University of Science
Viet Nam National University Ho Chi Minh City
Ho Chi Minh, Vietnam

Stem Cells in Clinical Applications brings some of the field's most renowned scientists and clinicians together with emerging talents and disseminates their cutting-edge clinical research to help shape future therapies. While each book tends to focus on regenerative medicine for a certain organ or system (e.g. liver, lung and heart; brain and spinal cord, etc.) each volume also deals with topics like the safety of stem cell transplantation, evidence for clinical applications, including effects and side effects, guidelines for clinical stem cell manipulation, and much more. Volumes will also discuss mesenchymal stem cell transplantation in autoimmune disease treatment, stem cell gene therapy in pre-clinical and clinical contexts, clinical use of stem cells in neurological degenerative disease, and best practices for manufacturers in stem cell production. Later volumes will be devoted to safety, ethics and regulations, stem cell banking, treatment of cancer, and genetic disease. This series provides insight not only into novel research in stem cells but also their clinical and real-world contexts. Each book in Stem Cells in Clinical Applications is an invaluable resource for advanced undergraduate students, graduate students, researchers and clinicians in the stem cell, tissue engineering, biomedical engineering, and regnerative medicine fields.

Phuc Van Pham • Sharmila Fagoonee
Editors

Stem Cell-Derived Exosomes and Microvesicles

Clinical Applications

Springer

Editors
Phuc Van Pham
VNUHCM-US Stem Cell Institute
University of Science
Viet Nam National University Ho Chi Minh City
Ho Chi Minh, Vietnam

Sharmila Fagoonee
c/o Molecular Biotechnology Center
Institute of Biostructure and Bioimaging, CNR
Turin, Italy

ISSN 2365-4198 ISSN 2365-4201 (electronic)
Stem Cells in Clinical Applications
ISBN 978-3-031-90697-8 ISBN 978-3-031-90698-5 (eBook)
https://doi.org/10.1007/978-3-031-90698-5

This Springer imprint is published by the registered company Springer Nature Switzerland AG
The registered company address is: Gewerbestrasse 11, 6330 Cham, Switzerland

Preface

The field of regenerative medicine has recently undergone remarkable transformations, with stem cell research at the forefront of these advancements. Among the most promising developments is the exploration of extracellular vesicles (EVs)—notably exosomes and microvesicles—originating from stem cells. These nanoscale entities serve as potent mediators of intercellular communication, adept at transferring proteins, lipids, and nucleic acids to target cells. The unique qualities of EVs have unlocked new possibilities for therapeutic applications, offering potential solutions to a myriad of diseases and medical challenges.

This book, *Stem Cell-Derived Exosomes and Microvesicles: Clinical Applications*, offers an exhaustive overview of the current landscape of research and clinical applications involving stem cell-derived EVs. Covering fundamental concepts and pioneering technologies, the chapters delve into the isolation, characterization, preservation, and functionalization of EVs, as well as their therapeutic potentials in diverse medical settings. Each chapter is crafted by distinguished experts, providing readers with profound insights into the latest advancements and future trajectories of this dynamic field.

We start with an introduction to extracellular vesicles, laying a robust foundation for understanding their biological significance and functional mechanisms. Following this, the book examines the technical intricacies of EV isolation and quality control, which are vital for their clinical efficacy and safety. Discussions extend to the preservation and manufacturing of EVs, spotlighting the challenges and innovations encountered in scaling up production for therapeutic purposes.

A notable section of the book is dedicated to the functionalization of EVs, investigating how engineering these vesicles can amplify their therapeutic effectiveness. The clinical applications of mesenchymal stem cell-derived EVs receive particular attention, focusing on their roles in addressing metabolic-dysfunction-associated steatotic liver disease (MASLD) and other conditions. Additionally, the potential of endothelial progenitor cell-derived EVs is scrutinized, bridging the divide between preclinical findings and clinical translation.

As the exploration of stem cell-derived EVs progresses, this book stands as a crucial resource for researchers, clinicians, and industry stakeholders aiming to harness the therapeutic prowess of these extraordinary vesicles. We aspire for the insights provided in these pages to inspire continued innovation and aid in the creation of novel therapies that enhance patient outcomes.

We express our sincere gratitude to all contributors for sharing their expertise and to our readers for embarking with us on this thrilling journey into the world of stem cell-derived exosomes and microvesicles.

Ho Chi Minh, Vietnam — Phuc Van Pham
Turin, Italy — Sharmila Fagoonee

Contents

Chapter 1
An Overview of Extracellular Vesicles

Sarah Tassinari and Cristina Grange

Abbreviations

APOs	Apoptotic bodies
EGFR	Epidermal growth factor receptor
ESCRT	Endosomal sorting complex required for transport
EVs	Extracellular vesicles
EXOs	Exosomes
HSP	Heat shock proteins
ILV	Intraluminal vesicles
lnRNAs	Long noncoding RNAs
LOs	Large oncosomes
miRNAs	microRNAs
MSCs	Mesenchymal stromal cells
MVB	Multivesicular body
MVs	Microvesicles
PEGs	Polyethylene glycol
Tsg101	Tumor Susceptibility gene 101

S. Tassinari
Department of Molecular Biotechnology and Health Sciences, University of Turin, Turin, Italy
e-mail: sarah.tassinari@unito.it

C. Grange (✉)
Department of Medical Sciences, University of Turin, Turin, Italy
e-mail: cristina.grange@unito.it

P. V. Pham, S. Fagoonee (eds.), *Stem Cell-Derived Exosomes and Microvesicles*, Stem Cells in Clinical Applications, https://doi.org/10.1007/978-3-031-90698-5_1

1 Introduction

New mechanisms of cell-to-cell communication have been discovered in recent decades, involving all soluble factors and bioactive molecules released by cells, collectively defined as the secretome. The secretome includes many classes of different molecules, such as hormones, chemokines, cytokines, extracellular proteinases, toxins, peptides, and extracellular vesicles (EVs) [33]. EVs are double-membrane vesicles shed by almost every type of cell, from prokaryotic to eukaryotic cells, including plants [79]. EVs are nano- to micrometer-scale spheres carrying bioactive proteins, lipids, and genetic materials. They contain several types of genetic material, such as mRNAs, microRNAs (miRNAs), long-noncoding RNAs (lnRNAs), circular RNAs, and DNAs, originating from parental cells [55, 60, 68].

EVs possess the distinctive property of communicating over long distances, having a role in the majority of physiological and pathological conditions [8, 54]. They display the capacity to transfer bioactive molecules and genetic information to target cells, influencing their function and behavior [3, 61].

In the last decade, the therapeutic potential of EVs and, in particular, of those released by stem cells has been clearly demonstrated for organ regeneration. Stem cell–derived EVs have been proposed as a new tool for regenerative medicine thanks to their activity on several aspects of regeneration and have been exploited in many preclinical and clinical trials [25, 56].

In this chapter, we will describe an overview of extracellular vesicles, focusing on their discovery, classification, and biogenesis. Moreover, present and future clinical applications of EVs are summarized. Finally, a brief discussion of innovative approaches for engineered EVs will also be considered.

2 History of EVs

In the last 20 years, the field of EVs has shown incredible success and has grown exponentially. The first intuition of what we now define EVs is linked to studies conducted by Chargaff and West in the 1940s. In a series of publications concerning the chemistry of blood coagulation, Chargaff initially observed a shortening in coagulation time due to high-speed plasma sedimentation [11]. Only in 1946, in a study conducted with Randolph West, a "particulate fraction" representing the first suggestion about these new particles was discovered [12]. Subsequently, after 17 years, Wolf provided the scientific community with the first image of what he called "platelet dust" [80], followed by Crawford in 1971, who also proposed the term "microparticles," showing some EV components such as lipids, ATP, and contractile proteins [16]. This latter innovative study was the first to identify the absence of cellular structures and highlighted their potential biological relevance. Between the late 1960s and early 1970s, numerous electron microscopy studies showed these submicron-sized structures in different cellular compartments: in the alveolar space

[66], in the matrix of hypertrophic cartilage [4, 7] or in the thyroid gland of bats [49]. In the same period, searching for virus-like particles in biological fluids, in accordance with the theory of the presence of a link between viral infections and cancer, a particulate was identified [40, 63]. However, it was not possible to correlate this particulate with viral infection due to its presence in both control and cancer fluids [18, 22, 28, 40, 42]. Toward the end of the 1960s, the scientific community was convinced of the idea that this particulate was not attributable to viruses but rather an artifact produced by the separation process [53]. Only in 1975, Dalton demonstrated that inside bovine serum, there were vesicles with similar characteristics to those of epithelial cells, dispelling any doubt about viral origin [17]. In the 1970s, several published studies demonstrated the presence of vesicular structures not only in mammals but also in algae, yeasts, and bacteria, beginning to reveal new cellular structures and promoting the existence of EVs [1, 13, 73]. The 1980s were marked by the birth of the field of EVs within the scientific community. Johnstone and Stahl described the release of intraluminal vesicles (ILVs) by cells and defined them as exosomes, thanks to their observation of the vesicular-mediated loss of transferrin receptors during reticulocyte maturation [50]. Harding later provided incredible electron microscopy images demonstrating the release of vesicles from the lumen of multivesicular bodies (MVBs) upon fusion with the plasma membrane [29]. From a conceptual point of view, Harding revealed the existence of a new intracellular sorting and trafficking pathway, today known as the exosomal secretion pathway. However, at that time, the role of EVs was not clear, and theories on their function as cellular waste or messengers remained a matter of debate.

The 1990s represented a turning point in the field of EVs, with confirmation of their existence and function. In 1991, Johnstone and coworkers demonstrated not only the presence of the transferrin receptor and nucleotide transporter in exosomes but also that cellular stresses involved internalization and release of these membrane components [35]. The diffusion of lipids and proteins on the vesicular membrane and the presence and function of flippase were also reported [24, 72]. Other studies revealed the main distinguishing features of these particles, such as the presence of tetraspanins, Rab, and ADP-ribosylation factor (ARF) [19, 71]. In recent years, an alteration in the number of vesicles has also been observed between physiological and pathological conditions. Thanks to this information, the field of EV application has expanded to several specific pathologies, such as cerebral ischemia, angina, and Crohn's disease [38, 52, 64]. Moreover, the main physical and biochemical characteristics begin to emerge, as well as evidence that the exosomes released by reticulocytes are enzymatically active [34]. Furthermore, it has also been found that vesicles from immune cells can present the antigen [59]. All this information opened the scientific community to the possibility of using EVs as an anticancer vaccine, a field of study followed by more than a decade of clinical trials [20]. The interest in EVs finally exploded in the twenty-first century along with the discovery, between 2006 and 2008, that EVs can functionally transfer RNAs [61, 65, 68]. This discovery has led the field of EVs to reach the maximum interest, allowing the study and identification of important mechanisms of EV release, regulation, and function for intercellular communication with potential implications in several aspects of the clinic, such as anticancer vaccines, biomarker discovery, and regenerative medicine.

3 Classification

Since EV identification, a wide variety of terms have been used to define these structures. As the field grew and the understanding of the different biogenesis pathways increased, it became clear that a distinct and precise nomenclature was needed to avoid confusion or misunderstanding. The scientific community created a global society of experts in the field of EVs called ISEV (https://www.isev.org), with the aim of generating guidelines for researchers and allowing discussion between scientists. The first guidelines, named Minimal Information for Studies of Extracellular Vesicles (MISEV), were defined in 2014 by the ISEV society, providing guidance for nomenclature and standardization of protocols and experiments [41]. According to the MISEV guidelines, all particles surrounded by a lipid bilayer, naturally released by a cell, not able to replicate, are defined as extracellular vesicles (Théry et al. [67] and more recently Welsh et al. [76]).

EVs can be categorized into different subclasses based on their size, density, biogenesis, or cellular origin. However, it is important to remark that there is no single, defining characteristic that can be used to definitively identify each type of EV. Furthermore, different EV subtypes often share similar properties in terms of size, marker expression, composition, or function. As a result, the nomenclature used to describe EVs is constantly evolving to reflect new scientific discoveries [76].

EVs can be classified based on some general characteristics:

1. Physical characteristics such as size: small EVs (<100 nm), medium EVs (<200 nm), and large EVs (>200 nm).
2. Intracellular origin, such as endosomal or ectosomal compartments.
3. Biochemical composition: CD63+/CD81+ EVs, Annexin A5-stained EVs.
4. Descriptions of characteristic or cell of origin (podocyte EVs, hypoxic EVs, large oncosomes, apoptotic bodies).
5. Function.

EVs are usually classified into different subtypes based on their size or secretion pathway (Fig. 1.1). The smaller vesicles are exosomes (EXOs) with a diameter of <120 nm. They originate from multivesicular bodies (MVBs) and can carry mRNAs and proteins involved in cell–cell signaling. They can be further classified into small EXOs (60–80 nm) and large EXOs (90–120 nm). Microvesicles (MVs) are generally larger than EXOs with a variable diameter of 100–1000 nm. They originate from the bubbling of the cell membrane when an increase in Ca^{2+} concentration occurs within the cell. This signal involves a lipid rearrangement inside the bilayer with the activation of enzymes such as flippase, translocase, and scramblase. MVs share some characteristic surface markers, such as CD63, with EXOs, and their function is similar since MVs transport cargo necessary for communication between

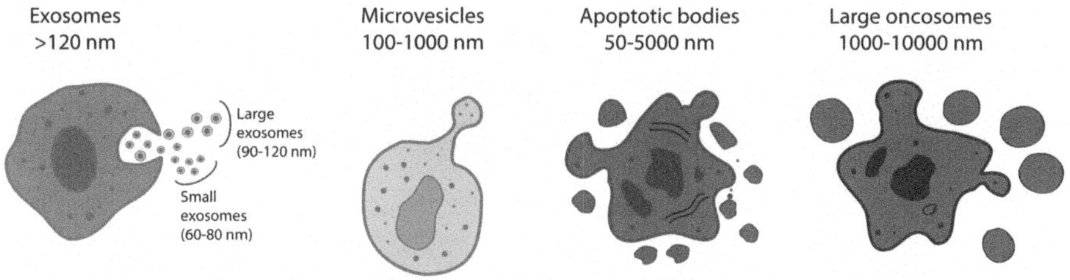

Fig. 1.1 EV classification. Schematic representation of the main EV types based on their dimension and origin

cells [14]. It has been shown that some EVs may arise during cell migration from specific cellular processes and they are called migrasomes [69]. Finally, cells undergoing apoptosis release a population of EVs called apoptotic bodies (APOs) through cell fragmentation and blebbing outwards (Fig. 1.1). The dimensions of these vesicles vary from 50 to 5000 nm, and the content is generally randomly packed. Populations of unique vesicles within a pathological context have also been identified, thereby presenting biophysical characteristics and specific composition. For example, oncosomes, or large oncosomes (LOs), are a population of EVs whose release appears to be uniquely circumscribed to cancer cells. They are large vesicles with a diameter of 1–10 μm and are capable of inducing migratory, invasive, and metastatic phenotypes. These vesicles originate from the shedding of the cell membrane of cancer cells presenting an amoeboid phenotype, a process induced by different stimuli such as overexpression of oncoproteins as well as activation of Akt1 or epidermal growth factor receptor (EGFR) pathways. Their formation is not linked to an increase in the rate of cell death, further emphasizing that LOs are a unique class and distinct from APOs [77]. The EVs that will be of interest for this book are mainly those of small size, possessing regenerative properties.

4 Biogenesis of Exosomes

Among the different types of EVs and their related origins, the biogenesis of exosomes is a process that has been better characterized [30, 58]. This process starts with a sequential invagination of the plasma membrane that generates the formation of intracellular multivesicular bodies (MVBs) that contain intraluminal vesicles (ILVs) (Fig. 1.2). The first invagination of the plasma membrane encloses a mixture of cell surface proteins and extracellular/soluble components. The progression of this process passes through the de novo formation of early-sorting endosomes and their maturation in late-sorting endosomes. The trans-Golgi network, together with the endoplasmic reticulum, contributes to the rearrangement and content of early-sorting endosomes [82]. This results in MVBs containing ILVs that are future exosomes. The fate of MVBs is now at a crossroad; they can fuse with the cellular

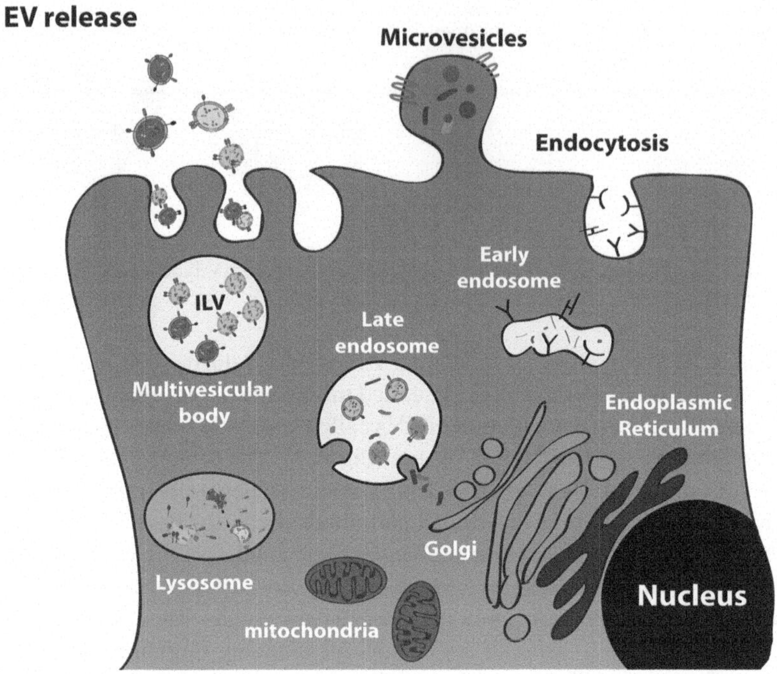

Fig. 1.2 EV biogenesis. Schematic representation of EV biogenesis, highlighting the formation of intraluminal vesicles, the maturation of EVs in the multivesicular body (MVB) and their release into the extracellular space

membrane to shed mature exosomes or with lysosomes or autophagosomes to be degraded [36, 82] (Fig. 1.2). The entire process is finely regulated by several proteins, and in particular, the endosomal sorting complex required for transport (ESCRT) has a central role in MVB formation and protein cargo sorting. ESCRT is composed of four separate proteins that cooperate [82]. Moreover, several other proteins participate in exosome maturation, such as tumor susceptibility gene 101 (TSG101), Ras-related protein GTPase, Rab, apoptosis-linked gene 2-interacting protein X (ALIX), and syndecan-1 in combination with ceramides, phospholipids, and sphingomyelinases. Ceramides are involved in membrane curvature and budding [45]. Other important proteins that contribute to this process are tetraspanins such as CD9, CD63, and CD9, which specialize in the compartmentalization of receptors and proteins in the plasma membrane of future exosomes [51]. Some of the most commonly expressed proteins in exosomes are listed in Table 1.1. The formation and maturation of exosomes have been described as alternative mechanisms to ESCRT-dependent mechanisms involving raft-based microdomains [82].

Apart from exosomes, other types of vesicles, such as MVs and APOs, originate from the budding of the plasma membrane of cells under normal conditions or during the late stage of apoptosis, respectively [82].

Table 1.1 Some of the common proteins in exosomes [6, 82]

Protein category	Specific protein as example
Tetraspanins	CD9, CD63, CD81, CD37, CD53, CD82
Heat shock proteins (HSP)	HSP70, HSP90, HSP27, HSP60, HSP84
Multivesicular body biogenesis	Tsg101, ALIX, Gag, Rab proteins
Cytoskeleton components	Actins, Erzin, Moesin, Myosin Cofilin-1, Tubulins, Vimentin, Advillin, Radixin
Cell adhesion	ICAM 1, Integrins, Lactadherin, MAC-1, CD166, ALCAM
Antigen presentation	Human leukocyte antigen class I and II
Trafficking and membrane fusion	Rab5,7, Rap1B, Annexins I, II, IV, V, VI Syntaxin-3, Synaptosomal-associated protein Dynamin

5 EVs and Intercellular Communication

EVs released by parental cells may interact with recipient cells, triggering changes in their phenotype, and several mechanisms have been proposed for EV uptake [44, 46]. This phenomenon is crucial for cell-to-cell communication not only under physiological conditions but also for the success of clinical/preclinical applications of EVs for tissue regeneration.

The specific mechanism depends on EV size and EV surface molecules. In particular, EVs may stimulate target cells by receptor–ligand interactions or may fuse with the plasma membrane and release their cargo in the cell cytoplasm. For EV internalization, several possibilities, which are not mutually exclusive, may occur, such as endocytosis mediated by clathrin or caveolin, cholesterol-rich lipid raft mechanisms, phagocytosis, macro- and micropinocytosis, or fusion of the plasma membrane (Fig. 1.3). Moreover, all the possible specific molecules important for EV binding to target cells have not been fully elucidated. There is a growing list of protein–protein interactions that facilitate EV attachment, favoring EV entrance into cells. The first binding can be mediated by adhesion proteins and by other classes of surface proteins, such as lectins, tetraspanins, heparan sulfate proteoglycans, and lipids [5, 9, 46]. In other cases, it has been described that cellular protrusions such as cilia or filopodia are the privileged sites for EV entry [31, 32].

Once EVs are internalized, their cargo is delivered and affects target cells [15, 46]. On the other hand, EVs can be immediately recycled in the extracellular space or degraded by lysosomes (Fig. 1.3). Regarding APOs, the presence of phosphatidylserine translocated to the outer leaflet of the lipid layer is important for their uptake in recipient cells [2].

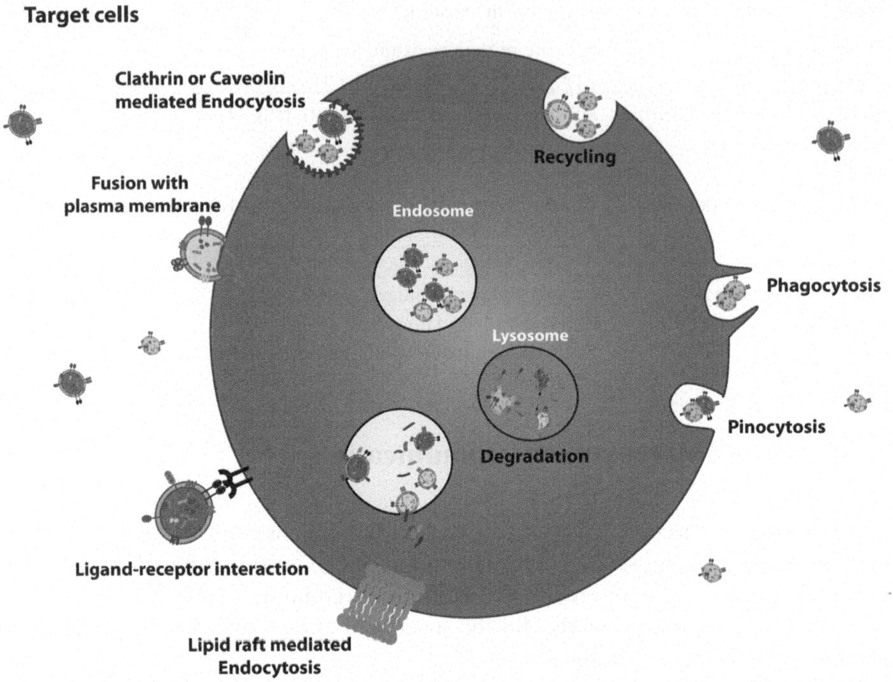

Fig. 1.3 EV uptake. After EV release in the intercellular space, they activate target cells through different mechanisms: endocytosis, fusion with the plasma membrane, or binding to cell surface receptors

6 EVs for Clinical Application: Advantages over Cell Therapy

Since EV discovery, its potential applications in several aspects of medicine appear immediately clear, and in recent decades, the number of publications reporting the therapeutic properties of EVs has increased enormously [62]. EVs, and in particular those released by stem cells, are now considered a promising cell-free therapy for the treatment of different pathologies.

Compared to cell therapy, the use of EVs presents many advantages. They display a higher safety profile as well as lower immunogenicity and, without the presence of a cell itself, the inability to maldifferentiate and the lack of replicating potential ([48], [25]). Moreover, EVs are biocompatible, biodegradable, and low cytotoxic, all important requirements for therapeutic applications [37, 47]. In addition, due to their small size and excellent biological tolerance, EVs may deliver their biological cargo to long-distance sites, cross biological barriers, and escape lung clearance [27, 81]. Of interest, another important feature of EVs to take into consideration is the capacity of targeting in terms of specific cell uptake and fast internalization [37]. Taken together, EVs are convincing candidates for regenerative medicine.

7 Cell Sources for the Production of Therapeutic EVs

The therapeutic potential of EVs depends on their content and characteristics, with the majority of them in common with the originating cells. Strong evidence has shown that stem cell–derived EVs recapitulate the pro-regenerative capacity of the stem cells per se, and in particular, those derived from mesenchymal stromal cells (MSCs) appear to be the optimal source to promote organ regeneration [26]. The classical definition of MSCs is "stromal cells that have the ability to self-renew and exhibit multilineage differentiation," and they can be isolated from different adult tissues, such as bone marrow, peripheral and cord blood, adipose tissue, or neonatal birth-associated tissues, including the placenta and umbilical cord [26]. In several preclinical models of renal, lung, liver, and heart injuries, it has been demonstrated that MSC EVs of different origins display equal or, in some cases, superior effects compared with MSCs [78]. EV administration in animal models favors a reduction in apoptosis, oxidative stress, and inflammation, and increases cell proliferation.

However, EVs with regenerative properties can have different origins apart from MSCs; they can be isolated from renal and cardiac progenitor cells, neural stem cells, and circulating endothelial progenitor cells and exploited for specific therapeutic applications. Moreover, the use of EVs isolated from embryonic stem cells and induced pluripotent stem cells has been investigated in preclinical injury models [74]. Details of some of these applications ARE better dissected in subsequent chapters.

8 Stem Cell–Derived EVs and Preclinical Application

Stem cell–derived EVs, and in particular MSC EVs, provide excellent benefits in several preclinical injury models; they have been successfully tested for cardiac, renal, liver, lung, nervous tissue, bone, muscle, and skeletal regeneration [23]. During organ damage, endogenous repair fails or is inadequate and slow; the administration of stem cell–derived EVs may prompt injured cells and damaged mitochondria to recover and may stimulate cell dedifferentiation and recruitment and activation of local stem cells or anti-inflammatory cells. On the other hand, EVs reduce apoptosis and oxidative stress and limit fibrosis [10, 39]. Apart from the regenerative potential in the mentioned organs, it has also been reported that stem cell–derived EVs may display beneficial effects in other diseases, such as lung, blood vessel, skin, and esophageal diseases [23].

It is important to mention the role of EVs in the treatment of inflammatory disease. Uncontrolled inflammation results in a wide range of chronic inflammatory diseases, such as rheumatoid arthritis, inflammatory bowel disease, and neurodegenerative disease. The current treatments are not satisfactory and can induce adverse effects and toxicity [75]. EVs mimic the role of cells inhibiting M1 macrophages and promoting M2 differentiation; similarly, EVs downregulate interleukin 6 (IL-6) production and increase the expression of IL-10 and TGF-β levels [75].

Another important application in which EVs may be used as therapeutic options is their application for wound healing. In fact, in vitro and in vivo studies and clinical trials showed the beneficial effects of stem cell–derived EVs. The wound healing process comprises complex sequential events such as angiogenesis, migration, extracellular matrix deposition, proliferation, and tissue remodeling [70]. EVs isolated from MSCs and human iPSCs promote fibroblast proliferation and migration and stimulate angiogenesis by activating Akt, ERK, and STAT3 in endothelial cells. Finally, adipose-derived MSCs have been shown to increase collagen and elastin synthesis [23].

Finally, MSC-EVs have recently been proposed as a new tool to improve graft survival due to their immunomodulatory, antifibrotic, and proangiogenic properties. The idea was to use EVs in a perfusion machine for preserving donor organs before transplant [26]. This application has been tested in preclinical studies for lung, kidney, and liver grafts and amelioration organ viability and functionality. The first clinical trial has been newly concluded on the use of MSC EVs in machine perfusion for marginal kidney. Kidneys treated with EVs appear with preserved renal architecture and less tubular damage [57].

9 Route of EV Administration and Delivery Strategies

The efficacy of EV treatment depends not only on EV regenerative cargo but also on specific technical factors, such as dose, regimen, and route of EV administration. In the majority of acute injury preclinical models, single EV administration is preferred, while in chronic settings, multiple EV doses are often investigated [25]. The route of EV administration depends on the type of tissue damage. The most commonly used for renal, liver, and cardiovascular models is the intravenous one, despite the rapid and well-known EV clearance by the liver. Local administration has the advantage of delivering high concentrations of EVs to the desired site, reducing off-target delivery. Intramuscular treatment is preferred for musculoskeletal conditions; subcutaneous treatment is indicated for cosmetics, while intrathecal treatment is exploited for neurodegenerative disease [48]. For wound and ulcers, topical or spray EV applications are preferred. Intranasal or inhalation administration is considered the easiest method and the most suitable for the clinic [21]. It has already been used successfully for the treatment of Parkinson's disease and COVID-19 [21].

New frontiers in EV research focus their efforts on the modulation of EV delivery to increase bioavailability and stability and improve their regenerative capacities. Some strategies are currently under investigation, such as the use of biomaterials to favor controlled EV release or engineering approaches to modify EV surface molecules [21]. In particular, it has already been reported that the encapsulation of EVs within different types of hydrogels generates a controlled and continuous EV release, with an improvement in EV efficacy [43]. Hydrogels appear to be a useful platform to prolong the EV existence and to design controlled releases.

Another approach that could be exploited to modulate EV uptake and biodistribution is the manipulation of EV surface molecules. This could be obtained with genetic and chemical manipulation, such as the use of polyethylene glycol (PEG), for example, generating chimeric and biomimetic EVs [21].

10 Conclusion and Future Perspectives

EVs, particularly those released by stem cells, possess great potential for regenerative medicine. The number of studies on the application of EVs in preclinical models of diseases is continuously increasing, and EVs are considered a promising approach for tissue regeneration. Their natural capacity to transfer biological messages among the fluid and to deliver their bioactive cargo to target cells can be exploited for therapeutic applications. As will be elucidated in subsequent chapters, there are several clinical trials going on, and few of them have already concluded with encouraging results. However, some key points for clinical translation, such as EV potency assays, standardization, and scaling up of EV production, remain to be defined.

On the other hand, knowledge in the EV field goes continuously further, and the next steps in EV applications are in progress. EVs could be utilized as nanocarriers for the transport of exogenous materials such as RNAs, proteins, or even drugs. Native therapeutically active EVs could be potentiated by genetic engineering, increasing the expression levels of their healing molecules or can be loaded with synthetic cargo. The chapters of this book will help to consolidate the knowledge in the field of EV and dissect all the aspects of their translational applications.

Acknowledgments We thank the ISEV and EVITA communities.

Conflict of Interest Authors declare no conflict of interest.

References

1. Aaronson S, Behrens U, Orner R, Haines TH. Ultrastructure of intracellular and extracellular vesicles, membranes, and myelin figures produced by Ochromonas danica. J Ultrastruct Res. 1971;35(5):418–30.
2. Akers JC, Gonda D, Kim R, Carter BS, Chen CC. Biogenesis of extracellular vesicles (EV): exosomes, microvesicles, retrovirus-like vesicles, and apoptotic bodies. J Neuro-Oncol. 2013;113(1):1–11.

3. Aliotta JM, Pereira M, Johnson KW, et al. Microvesicle entry into marrow cells mediates tissue-specific changes in mRNA by direct delivery of mRNA and induction of transcription. Exp Hematol. 2010;38(3):233–45. https://doi.org/10.1016/j.exphem.2010.01.002.
4. Anderson HC. Vesicles associated with calcification in the matrix of epiphyseal cartilage. J Cell Biol. 1969;41(1):59–72.
5. Barrès C, Blanc L, Bette-Bobillo P, et al. Galectin-5 is bound onto the surface of rat reticulocyte exosomes and modulates vesicle uptake by macrophages. Blood. 2010;115(3):696–705.
6. Beach A, Zhang HG, Ratajczak MZ, Kakar SS. Exosomes: an overview of biogenesis, composition and role in ovarian cancer. J Ovarian Res. 2014;7:14.
7. Bonucci E. Fine structure of early cartilage calcification. J Ultrastruct Res. 1967;20(1):33–50.
8. Borgovan T, Crawford L, Nwizu C, Quesenberry P. Stem cells and extracellular vesicles: biological regulators of physiology and disease. Am J Physiol Cell Physiol. 2019;317(2):C155–66.
9. Bruno S, Grange C, Deregibus MC, et al. Mesenchymal stem cell-derived microvesicles protect against acute tubular injury. J Am Soc Nephrol. 2009;20(5):1053–67.
10. Bruno S, Chiabotto G, Camussi G. Extracellular vesicles: a therapeutic option for liver fibrosis. Int J Mol Sci. 2020;21(12):4255.
11. Chargaff EJ. Cell structure and the problem of blood coagulation. J Biol Chem. 1945;160(1):351–9.
12. Chargaff E, West R. The biological significance of the thromboplastic protein of blood. J Biol Chem. 1946;166(1):189–97.
13. Chigaleichik AG, Belova LA, Grishchenko VM, Rylkin SS. Several properties of the extracellular vesicles of Candida tropicalis yeasts grown on n-alkanes. Mikrobiologiia. 1977;46(3):467–71.
14. Chuo STY, Chien JCY, Lai CPK. Imaging extracellular vesicles: current and emerging methods. J Biomed Sci. 2018;25:91.
15. Corbeil D, Santos MF, Karbanová J, Kurth T, Rappa G, Lorico A. Uptake and fate of extracellular membrane vesicles: nucleoplasmic reticulum-associated late endosomes as a new gate to intercellular communication. Cells. 2020;9(9):1931.
16. Crawford N. The presence of contractile proteins in platelet microparticles isolated from human and animal platelet-free plasma. Br J Hematol. 1971;21(1):53–69.
17. Dalton AJ. Microvesicles and vesicles of multivesicular bodies versus "virus-like" particles. J Natl Cancer Inst. 1975;54(5):1137–48.
18. Dmochowski L, Langford PL, Williams WC, Liebelt AG, Liebelt RA. Electronmicroscopic and bioassay studies of milk from mice of high and low mammary-cancer and high and low leukemia strains. J Natl Cancer Inst. 1968;40(6):1339–58.
19. Escola J-M, Kleijmeer MJ, Stoorvogel W, Griffith JM, Yoshie O, Geuze HJ. Selective enrichment of tetraspan proteins on the internal vesicles of multivesicular endosomes and on exosomes secreted by human B-lymphocytes. J Biol Chem. 1998;273(32):20121–7.
20. Escudier B, Dorval T, Chaput N, et al. Vaccination of metastatic melanoma patients with autologous dendritic cell (DC)-derived exosomes: Results of the first phase I clinical trial. J Transl Med. 2005;3(1):10.
21. Esmaeili A, Alini M, Baghaban Eslaminejad M, Hosseini S. Engineering strategies for customizing extracellular vesicle uptake in a therapeutic context. Stem Cell Res Ther. 2022;13(1):129.
22. Fawcett DW. Electron microscope observations on intracellular virus-like particles associated with the cells of the Lucke renal adenocarcinoma. J Biophys Biochem Cytol. 1956;2(6):725–42.
23. Fuloria S, Subramaniyan V, Dahiya R, et al. Mesenchymal stem cell-derived extracellular vesicles: regenerative potential and challenges. Biology (Basel). 2021;10(3):172.
24. Gawrisch K, Stibenz D, Möps A, Arnold K, Linss W, Halbhuber K-J. The rate of lateral diffusion of phospholipids in erythrocyte microvesicles. Biochim Biophys Acta. 1986;856(3):443–7.
25. Grange C, Skovronova R, Marabese F, Bussolati B. Stem cell-derived extracellular vesicles and kidney regeneration. Cells. 2019;8(10):1240.
26. Grange C, Bellucci L, Bussolati B, Ranghino A. Potential applications of extracellular vesicles in solid organ transplantation. Cells. 2020;9(2):369.

27. Guo Y, Wang H, Huang L, et al. Small extracellular vesicles-based cell-free strategies for therapy. MedComm (2020). 2021;2(1):17–26.
28. Haguenau F. Cancer of the breast in the female. Comparative electron microscope and optical microscope study. Bull De L Assoc Fr Pour L Etude Du Cancer. 1959;46:177–211.
29. Harding C, Heuser J, Stahl P. Receptor-mediated endocytosis of transferrin and recycling of the transferrin receptor in rat reticulocytes. J Cell Biol. 1983;97(2):329–39.
30. Hessvik NP, Llorente A. Current knowledge on exosome biogenesis and release. Cell Mol Life Sci. 2018;75(2):193–208.
31. Heusermann W, Hean J, Trojer D, et al. Exosomes surf on filopodia to enter cells at endocytic hot spots, traffic within endosomes, and are targeted to the ER. J Cell Biol. 2016;213(2):173–84.
32. Hogan MC, Manganelli L, Woollard JR, et al. Characterization of PKD protein-positive exosome-like vesicles. J Am Soc Nephrol. 2009;20(2):278–88.
33. Huang Y, Li X, Yang L. Hydrogel encapsulation: taking the therapy of mesenchymal stem cells and their derived secretome to the next level. Front Bioeng Biotechnol. 2022;10:859927. Published 2022 Apr 1
34. Johnstone RM, Bianchini A, Teng K. Reticulocyte maturation and exosome release: transferrin receptor containing exosomes shows multiple plasma membrane functions. Blood. 1989;74(5):1844–51.
35. Johnstone RM, Mathew A, Mason AB, Teng K. Exosome formation during maturation of mammalian and avian reticulocytes: evidence that exosome release is a major route for externalization of obsolete membrane proteins. J Cell Physiol. 1991;147(1):27–36.
36. Kalluri R, LeBleu VS. The biology, function, and biomedical applications of exosomes. Science. 2020;367(6478):eaau6977.
37. Kooijmans SAA, Schiffelers RM, Zarovni N, Vago R. Modulation of tissue tropism and biological activity of exosomes and other extracellular vesicles: New nanotools for cancer treatment. Pharmacol Res. 2016;111:487–500.
38. Lee YJ, Jy W, Horstman LL, Janania J, Reyes Y, Kelley RE, Ahn YS. Elevated platelet microparticles in transient ischemic attacks, lacunar infarcts, and multiinfarct dementias. Thromb Res. 1993;72(4):295–304.
39. Lelek J, Zuba-Surma EK. Perspectives for future use of extracellular vesicles from umbilical cord- and adipose tissue-derived mesenchymal stem/stromal cells in regenerative therapies-synthetic review. Int J Mol Sci. 2020;21(3):799.
40. Levine PH, Horoszewicz JS, Grace JT, Chai LS, Ellison RR, Holland JF. Relationship between clinical status of leukemic patients and virus-like particles in their plasma. Cancer. 1967;20(10):1563–77.
41. Lötvall J, Hill AF, Hochberg F, et al. Minimal experimental requirements for definition of extracellular vesicles and their functions: a position statement from the International Society for Extracellular Vesicles. J Extracell Vesicles. 2014;3:26913.
42. Lunger PD, Lucas JC, Shipkey FH. The ultramorphology of milk fractions for normal and breast cancer patients. A preliminary report. Cancer. 1964;17:549–57.
43. Mardpour S, Ghanian MH, Sadeghi-Abandansari H, Mardpour S, Nazari A, Shekari F, Baharvand H. Hydrogel-mediated sustained systemic delivery of mesenchymal stem cell-derived extracellular vesicles improves hepatic regeneration in chronic liver failure. ACS Appl Mater Interfaces. 2019;11(41):37421–33.
44. Mathieu M, Martin-Jaular L, Lavieu G, Théry C. Specificities of secretion and uptake of exosomes and other extracellular vesicles for cell-to-cell communication. Nat Cell Biol. 2019;21(1):9–17.
45. Matsui T, Osaki F, Hiragi S, Sakamaki Y, Fukuda M. ALIX and ceramide differentially control polarized small extracellular vesicle release from epithelial cells. EMBO Rep. 2021;22(5):e51475.
46. Mulcahy LA, Pink RC, Carter DR. Routes and mechanisms of extracellular vesicle uptake. J Extracell Vesicles. 2014;3 https://doi.org/10.3402/jev.v3.24641.

47. Murphy DE, de Jong OG, Brouwer M, et al. Extracellular vesicle-based therapeutics: natural versus engineered targeting and trafficking. Exp Mol Med. 2019;51(3):1–12.
48. Muthu S, Bapat A, Jain R, Jeyaraman N, Jeyaraman M. Exosomal therapy – a new frontier in regenerative medicine. Stem Cell Investig. 2021;8:7.
49. Nunez EA, Wallis J, Gershon MD. Secretory processes in follicular cells of the bat thyroid. 3. The occurrence of extracellular vesicles and colloid droplets during arousal from hibernation. Am J Anat. 1974 Oct;141(2):179-201. doi: 10.1002/aja.1001410203. PMID: 4415703.
50. Pan B-T, Johnstone RM. Fate of the transferrin receptor during maturation of sheep reticulocytes in vitro: selective externalization of the receptor. Cell. 1983;33(3):967–78.
51. Perez-Hernandez D, Gutierrez-Vazquez C, Jorge I, Lopez-Martin S, Ursa A, Sanchez-Madrid F, Vazquez J, Yanez-Mo M. The intracellular interactome of tetraspanin-enriched microdomains reveals their function as sorting machineries toward exosomes. J Biol Chem. 2013;288(17):11649–61.
52. Powell JJ, Harvey RS, Thompson RP. Microparticles in Crohn's disease–has the dust settled? Gut. 1996;39(2):340–1.
53. Prince AM, Adams WR. Virus-like particles in human plasma and serum: role of platelet lysosomes. J Natl Cancer Inst. 1966;37(2):153–66.
54. Quesenberry P, Goldberg LR. A new stem cell biology: transplantation and baseline, cell cycle and exosomes. Adv Exp Med Biol. 2018;1056:3–9.
55. Quesenberry PJ, Aliotta J, Deregibus MC, Camussi G. Role of extracellular RNA-carrying vesicles in cell differentiation and reprogramming. Stem Cell Res Ther. 2015;6:153.
56. Racchetti G, Meldolesi J. Extracellular vesicles of mesenchymal stem cells: therapeutic properties discovered with extraordinary success. Biomedicines. 2021;9(6):667.
57. Rampino T, et al. Extracellular vesicles derived from mesenchymal stromal cells delivered during hypothermic oxygenated machine perfusion repair ischemic/reperfusion damage of kidneys from extended criteria donors. Biology. 2022;11:350.
58. Raposo G, Stoorvogel W. Extracellular vesicles: exosomes, microvesicles, and friends. J Cell Biol. 2013;200(4):373–83.
59. Raposo G, Nijman HW, Stoorvogel W, Liejendekker R, Harding CV, Melief CJ, Geuze HJ. B lymphocytes secrete antigen-presenting vesicles. J Exp Med. 1996;183(3):1161–72.
60. Ratajczak MZ, Ratajczak J. Horizontal transfer of RNA and proteins between cells by extracellular microvesicles: 14 years later. Clin Transl Med. 2016;5(1):7.
61. Ratajczak J, Miekus K, Kucia M, et al. Embryonic stem cell-derived microvesicles reprogram hematopoietic progenitors: evidence for horizontal transfer of mRNA and protein delivery. Leukemia. 2006;20(5):847–56.
62. Sanmartin MC, Borzone FR, Giorello MB, Yannarelli G, Chasseing NA. Mesenchymal stromal cell-derived extracellular vesicles as biological carriers for drug delivery in cancer therapy. Front Bioeng Biotechnol. 2022;10:882545.
63. Seman G, Gallager HS, Lukeman JM, Dmochowski L. Studies on the presence of particles resembling RNA virus particles in human breast tumors, pleural effusions, their tissue cultures, and milk. Cancer. 1971;28(6):1431–42.
64. Singh N, Gemmell CH, Daly PA, Yeo EL. Elevated platelet-derived microparticle levels during unstable angina. Can J Cardiol. 1995;11(11):1015–21.
65. Skog J, Würdinger T, VanRijn S, Meijer DH, Gainche L, Curry WT, Carter BS, Krichevsky AM, Breakefield XO. Glioblastoma microvesicles transport RNA and proteins that promote tumor growth and provide diagnostic biomarkers. Nat Cell Biol. 2008;10(12):1470–6.
66. Sun CN. Lattice structures and osmiophilic bodies in the developing respiratory tissue of rats. J Ultrastruct Res. 1966;15(3):380–8.
67. Théry C, et al. Minimal information for studies of extracellular vesicles 2018 (MISEV2018): a position statement of the International Society for Extracellular Vesicles and update of the MISEV2014 guidelines. J Extracell Vesicles. 2018;7(1):1535750.

68. Valadi H, Ekström K, Bossios A, Sjöstrand M, Lee JJ, Lötvall JO. Exosome-mediated transfer of mRNAs and microRNAs is a novel mechanism of genetic exchange between cells. Nat Cell Biol. 2007;9:654–9.
69. Van Niel G, Carter DRF, Clayton A, Lambert DW, Raposo G, Vader P. Challenges and directions in studying cell-cell communication by extracellular vesicles. Nat Rev Mol Cell Biol. 2022;23:369–82. https://doi.org/10.1038/s41580-022-00460-3.
70. Velnar T, Bailey T, Smrkolj V. The wound healing process: an overview of the cellular and molecular mechanisms. J Int Med Res. 2009;37(5):1528–42.
71. Vidal MJ, Stahl PD. The small GTP-binding proteins Rab4 and ARF are associated with released exosomes during reticulocyte maturation. Eur J Cell Biol. 1993;60(2):261–7.
72. Vidal M, Sainte-Marie J, Philippot JR, Bienvenue A. Asymmetric distribution of phospholipids in the membrane of vesicles released during in vitro maturation of Guinea pig reticulocytes: Evidence precluding a role for aminophospholipid translocase. J Cell Physiol. 1989;140(3):455–62.
73. Vysotskii VV, Mazurova IK, Shmeleva EA. Extracellular material of some representatives of the genus Corynebacterium (the electron microscopic aspect). Zh Mikrobiol Epidemiol Immunobiol. 1977;8:90–5.
74. Wang AYL. Human induced pluripotent stem cell-derived exosomes as a new therapeutic strategy for various diseases. Int J Mol Sci. 2021;22(4):1769.
75. Wang S, Lei B, Zhang E, et al. Targeted therapy for inflammatory diseases with mesenchymal stem cells and their derived exosomes: from basic to clinics. Int J Nanomedicine. 2022;17:1757–81.
76. Welsh JA, et al. Minimal information for studies of extracellular vesicles (MISEV2023): From basic to advanced approaches. J Extracell Vesicles. 2024;13(2):e12404.
77. Willms E, Cabañas C, Mäger I, Wood M, Vader P. Extracellular vesicle heterogeneity: subpopulations, isolation techniques, and diverse functions in cancer progression. Front Immunol. 2018;9:738.
78. Witwer KW, Van Balkom BWM, Bruno S, et al. Defining mesenchymal stromal cell (MSC)-derived small extracellular vesicles for therapeutic applications. J Extracell Vesicles. 2019;8(1):1609206.
79. Woith E, Fuhrmann G, Melzig MF. Extracellular vesicles-connecting kingdoms. Int J Mol Sci. 2019;20(22):5695.
80. Wolf P. The nature and significance of platelet products in human plasma. Br J Hematol. 1967;13(3):269–88.
81. Yang T, Martin P, Fogarty B, et al. Exosome delivered anticancer drugs across the blood–brain barrier for brain cancer therapy in Danio rerio. Pharm Res. 2015;32(6):2003–14.
82. Zhang Y, Liu Y, Liu H, Tang WH. Exosomes: biogenesis, biologic function and clinical potential. Cell Biosci. 2019;9:19.

Chapter 2
Isolation of Exosomes from Cell Culture Supernatants

Ngoc Bich Vu

1 Introduction

Exosomes are small extracellular vesicles (30–150 nm in diameter) that play crucial roles in cell communication, carrying proteins, lipids, and nucleic acids from their cells of origin to other cells. They are involved in various physiological and pathological processes, making their isolation important for both research and clinical diagnostics. Exosome isolation is a fundamental method in the field of exosome technology. Multiple methods are available for isolating exosomes from the liquid portion of stem cells. While various procedures can be employed to isolate exosomes, certain techniques are capable of yielding highly pure exosomes with a high recovery rate. Some of these techniques can be scaled up for isolation of exosomes for therapeutic purposes. The utilization of exosome isolation strategies is contingent upon certain factors: (1) What is the maximum number of exosomes that can be isolated using this technique? (2) What is the recovery rate of the technique? (3) Is the isolated exosome pure and acceptable for therapeutic applications?

Due to the significant potential of exosomes for research and applications, numerous approaches have been developed for exosome isolation. This chapter aims to categorize various technologies into two distinct groups: conventional and emerging technologies. Most technologies rely on the size and density of exosomes relative to other particles in the supernatant. Conventional methods use centrifugation or filtration to isolate exosomes [1–5], whereas developing

N. B. Vu (✉)
Laboratory of Stem Cell Research and Application, University of Science, Viet Nam National University, Ho Chi Minh City, Viet Nam

VNUHCM-US Stem Cell Institute, University of Science, Viet Nam National University Ho Chi Minh City, Ho Chi Minh, Vietnam
e-mail: vbngoc@hcmus.edu.vn

P. V. Pham, S. Fagoonee (eds.), *Stem Cell-Derived Exosomes and Microvesicles*, Stem Cells in Clinical Applications, https://doi.org/10.1007/978-3-031-90698-5_2

Table 2.1 Some techniques for exosome isolation

Method	Principle	Advantages	Limitations
Ultracentrifugation	Size and density separation	Gold standard, no reagents needed	Time-consuming, low purity
Size exclusion chromatography	Size-based separation	High purity, preserves integrity	Limited sample volume, equipment needed
Polymer precipitation	Solubility reduction	Simple, fast, no equipment needed	Low purity, alters the exosome structure
Immunoaffinity capture	Antibody-based targeting	High specificity, cell-type specific	Expensive, low yield
Microfluidics	Size/density/immunoaffinity	High sensitivity, small sample volume	Specialized equipment, low throughput
Tangential flow filtration	Continuous size separation	High yield, scalable	Costly, requires additional purification

technologies utilize microfluidics for exosome separation [6–8]. Conventional exosome isolation techniques encompass a range of methods, including differential centrifugation, density gradient ultracentrifugation, precipitation-based approaches, immunoaffinity, ultrafiltration, and size exclusion chromatography (Table 2.1).

2 Conventional Techniques

The selection of an exosome separation methodology is contingent upon the specific demands of the subsequent study or application, including the necessity for either high purity or high yield, as well as the accessibility of equipment and reagents (Fig. 2.1).

2.1 Differential Centrifugation

Differential centrifugation (DC) is widely recognized as the leading and most inclusive technique for isolating exosomes. It uses a series of centrifugation speeds, ranging from low to high, to efficiently separate cells, cellular debris, proteins, and vesicles [9–11].

The centrifugation process for this method begins with a force ranging from 300 to 2000×*g*. This stage is crucial for separating cells, apoptotic bodies, and cell debris, which precipitate as dense pellets [12–14]. Next, a strategic rise in centrifugal force to a range of 5000 to 10,000×*g* separates large extracellular vesicles (EVs) and protein pellets, which are distinguished by their distinct size and density properties. The procedure concludes by subjecting the supernatant to a centrifugal force of 100,000×*g* for 1 to 3 h to separate exosomes. Each stage of the centrifugation

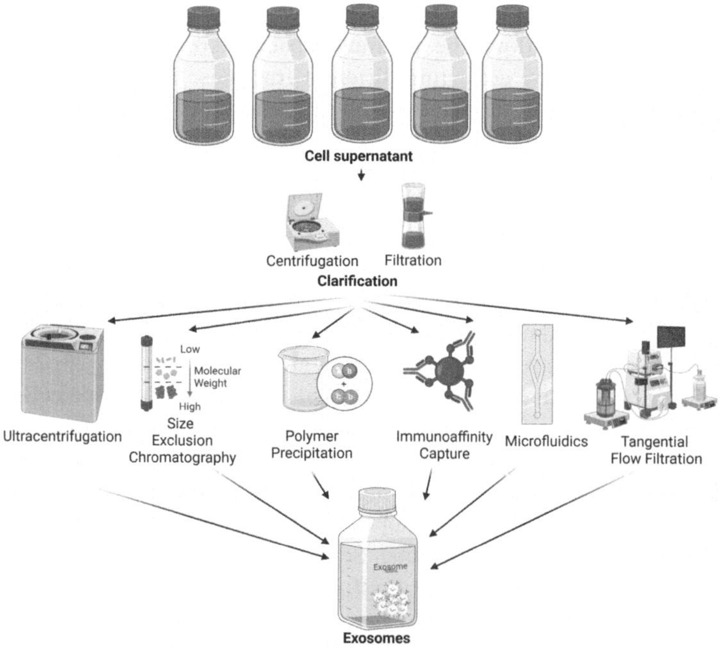

Fig. 2.1 Various common methods used for isolating exosomes from cell culture media include ultracentrifugation, size exclusion chromatography, polymer precipitation, immunoaffinity capture, microfluidics, and tangential flow filtration. Each method utilizes distinct separation principles, resulting in unique advantages and disadvantages. (The figure is created using biorender.com)

procedure is carried out with great care at a regulated temperature of 4 °C. In the final step, the pellets are cleansed and reconstituted in Phosphate buffer saline (PBS) prior to the last round of centrifugation at 100,000×*g* for 1 h [12–15].

This process of isolating exosomes using differential centrifugation is a relatively straightforward and cost-effective procedure [16, 17], with the exception of the initial equipment expenditure. Nevertheless, this procedure has certain disadvantages, especially for scalable manufacturing [17]. Cellular debris and apoptotic bodies may taint the exosomes during centrifugation [16]. In addition, these procedures are time-consuming and require specialized equipment [17].

2.2 *Density Gradient Ultracentrifugation*

Density gradient centrifugation (DG UC) improves upon the conventional DC method for isolating exosomes by gradually decreasing the density of the medium from the bottom to the top. This approach replicates the numerous steps of DC by placing the sample into a gradient medium that separates into layers with different densities. When ultracentrifugation is started with a specific force, particles in the sample move until they reach a layer where their density matches that of the surrounding medium. This allows the accurate separation of exosomes from vesicles with different densities by collecting different fractions. This technique reports a higher purity of exosomes compared to conventional centrifugation [18–20].

Sucrose and iodixanol (OptiPrep) are commonly selected for these noncontinuous gradients because of their effectiveness and extensive documentation in academic literature [5, 21, 22]. Although sucrose is attractive due to its nontoxic nature, neutral pH, and ability to produce pure exosomes, iodixanol is preferred in multiple studies due to its advantages in preventing vesicle damage and sluggish sedimentation caused by thick, highly concentrated sucrose solutions. Iodixanol's osmotic properties help to maintain a balanced solution at different densities, helping to preserve the shape and size of vesicles during gradient navigation [23, 24].

Before DG UC, initial procedures on the cell supernatant exclude cells and bigger fragments and employing a sequence of centrifugation or filtration procedures. Initially, the sample is subjected to low-speed centrifugation to separate and collect cells. Subsequently, centrifugation at higher speeds separates and collects larger particles and vesicles, resulting in a supernatant that contains exosomes. The density gradient is prepared within the tube. Several forms of gradients that can be utilized include sucrose, iodixanol (OptiPrep), or percoll. Gradients in the tube can be either continuous, with a gradual increase in density from top to bottom, or discontinuous, with discrete layers of various densities. The supernatant containing exosomes is carefully placed on the density gradient, and the tube is inserted into an ultracentrifuge and subjected to rapid rotation at elevated velocities (usually 100,000×*g* or above) for several hours [25]. During this period, particles inside the sample move through the gradient to reach a point where their density matches that of the surrounding gradient layer, thus separating them based on their density. Following ultracentrifugation, the gradient is divided into fractions, each of which may contain exosomes. The fractions are sequentially collected from the upper to lower part of the tube.

The DG UC method is well regarded for its capacity to produce exosomes of exceptional purity, establishing it as the benchmark for exosome research [26]. Nevertheless, this process is time-consuming, requires specialized equipment, and has comparatively low throughput compared to alternative techniques such as size exclusion chromatography or immunoaffinity capture. Furthermore, the effectiveness of exosome retrieval and the preservation of the separated exosomes might be affected by the composition of the density gradient and the parameters of ultracentrifugation. Although DG UC has several limitations, it continues to be a valuable tool in the field of exosome research, enabling valuable insights into the extensive potential of these vesicles for diagnostics and therapeutic purposes [27, 28].

2.3 Size Exclusion Chromatography

Size exclusion chromatography (SEC), commonly referred to as gel filtering, is a method of separating particles by their size. It utilizes a stationary phase composed of polymeric porous beads with defined pore diameters. This technique facilitates the categorization of particles based on their radii, with smaller particles experiencing a delay as they enter the pores, while larger particles bypass the pores and move more swiftly between the beads [29]. Notable commercial products for purifying exosomes using SEC include qEV columns manufactured by iZON Science, PURE-EVs produced by Hansa Biomed, and EVSecond purification columns offered by GL Sciences. The Exo-spin, developed by Cell Guidance Systems, is a hybrid technology that combines SEC with precipitation.

Samples are often cleared before SEC exosome separation. Cells, cell debris, and large vesicles or aggregates that could interfere with column separation are removed by centrifugation or filtration. In an SEC column appropriate for exosome isolation, porous beads segregate particles by size in columns. Pores should let exosomes through but reject and elute bigger entities first. Usually PBS or another buffer that preserves exosomes is used to equilibrate the column and prepare it for sample application; exosomes will not adhere to the column matrix after this phase. To prevent disturbing the packed bed, the pre-cleared sample is gently applied to the column, which is eluted using an equilibration buffer. The column's void volume—the volume excluded from the beads' pores—elutes exosomes. Knowledge of the SEC setup's expected elution profile enables accurate extraction of exosome-containing fractions.

Exosome isolation via SEC offers benefits: SEC gently changes exosomes' physical and biological features [30, 31]. Studies on exosome function and therapeutic usage require this. SEC isolates exosomes from other tiny particles and proteins, making downstream analytical and biomolecular studies easier [32]. SEC requires only a normal chromatography apparatus, unlike ultracentrifugation, making it accessible to most labs [29]. SEC may be scaled up or down for different sample volumes, making it useful for small-scale research and bioprocessing [33]. SEC does not require harmful reagents that could damage exosomes or endanger operators. Potential drawbacks include lower production of exosomes than with ultracentrifugation [32], limiting high-yield applications. SEC processes fewer volumes in each run than ultrafiltration or ultracentrifugation, making it time-consuming to process large sample volumes [34]. The SEC column may elute exosomes in a higher buffer volume than the sample, requiring a concentration step before analysis. Frequent use of biological material can contaminate and degrade columns, leading to increased operational costs from column repair and replacement. Size-based SEC separates particles but may not differentiate exosomes from protein clumps or lipoproteins that co-elute with them.

2.4 Anion Exchange Chromatography

Anion exchange chromatography (AEX) for exosome isolation is based on the interaction between the negatively charged surfaces of exosomes and the positively charged groups on the surface of the anion exchange resin. This technique utilizes the charge characteristics of exosomes to achieve the separation from other constituents in a sample [35]. At normal pH, exosomes have an overall negative surface charge due to sialic acids and other anionic groups on glycoproteins and glycolipids, as well as a diversity of membrane proteins, lipids, and other compounds on their surface. This makes exosomes capable of interacting with positively charged molecules under suitable conditions. The resin utilized in AEX is composed of a matrix, often consisting of agarose, sepharose, or polymer-based beads, to which positively charged groups ($-NH_3^+$, $-NH_2^+$) are chemically bonded. The charged groups can engage with negatively charged substances in a solution, such as exosomes, through electrostatic attractions [36, 37].

The AEX technique for exosome isolation is straightforward. A sample containing exosomes is introduced onto an anion exchange column, where the exosomes that have a negative charge are drawn toward and adhere to the positively charged sites on the resin. The conditions are adjusted to maximize the pH and ionic strength of the buffer to enhance the specific binding of exosomes while minimizing the binding of other negatively charged particles or biomolecules present in the sample. Following the binding, a buffer solution eliminates any weakly attached or unattached substances from the column, therefore decreasing nonspecific background and enhancing the quality of the exosomes. Optimal ionic strength and pH of the washing buffer are essential: sufficiently potent to eliminate undesired substances, while avoiding excessive strength that could cause the elution of exosomes. To separate the exosomes from the column, the buffer's ionic strength is raised (or the pH is modified) to disturb the electrostatic connections between the exosomes and the resin. This can be accomplished by utilizing a buffer solution with an elevated concentration of salt or by employing a salt gradient. With an increase in ionic strength, the electrostatic forces become less effective due to shielding, resulting in the release of exosomes into the eluate.

AEX has high scalability, allowing expansion to accommodate tiny research samples as well as larger volumes needed for medicinal applications [38, 39]. AEX can achieve specificity toward exosomes over other vesicles or particles of comparable size by precise adjustment of conditions, taking into account their charge characteristics [38]. Nevertheless, this strategy has some drawbacks, such as the intricacy of preparing samples, the possibility of nonspecific binding, and difficulties associated with exosome population variability [35]. It is important to take these constraints into account when selecting this method for exosome isolation. In many cases, it can be advantageous to combine AEX with other techniques to overcome these problems and reach the appropriate level of purity and yield of exosomes.

2.5 *Ultrafiltration (Tangential Flow Filtration or TFF)*

Ultrafiltration (UF) is a membrane-based filtration process used in the separation, concentration, and purification of molecules, including the isolation of exosomes from various biological fluids like blood, urine, and cell culture media. This technique exploits size exclusion or sieving principles to separate molecules based on their size and, to a lesser extent, their shape. Ultrafiltration involves the use of a semi-permeable membrane with specific pore sizes that allow smaller molecules and ions to pass through while retaining larger molecules or particles. The process depends on the ability of the membrane to retain molecules that are larger than its pore size, such as exosomes, while allowing smaller molecules like proteins, salts, and other tiny solutes to pass through. The selective retention of exosomes is a result of size exclusion, which enables the concentration of exosomes in the retentate (the portion that does not pass through the membrane), while other components are removed in the permeate (the portion that passes through the membrane). The driving force for the separation process is a pressure gradient across the membrane, leading to the convective transport of solutes.

The UF method of tangential flow filtration (TFF), also known as cross-flow filtration, is a filtration process where the feed stream flows tangentially across the surface of the filter membrane, rather than perpendicularly. TFF is an efficient method for exosome isolation.

Careful choice of membrane pore size is crucial to effectively separate the exosomes using TFF. Exosome dimensions range from 30 to 150 nm. The use of a membrane with a suitable molecular weight cut-off (MWCO) can efficiently concentrate exosomes while separating them from smaller molecules and particles. As most MWCO values are expressed in kDa (kilo dalton), it is crucial to have a clear understanding of the membrane pore size. A differential pressure across the membrane, achieved using mechanical means, such as a pump, or applying centrifugal force, enables the transport of solutes. Optimizing the pressure is necessary to achieve efficient separation, while limiting harm to the exosomes and avoiding membrane blockage or fouling.

UF or TFF offers several benefits for isolating exosomes. UF systems are very adaptable and can efficiently handle different sample volumes, ranging from small laboratory samples to larger amounts needed for clinical applications. When compared to alternative procedures such as ultracentrifugation or density gradient centrifugation, UF is faster, particularly during the early concentration steps [40]. UF functions under moderate circumstances, thereby preserving the structural and functional properties of isolated exosomes [40]. It is a versatile technique that may be used with many source materials. It can also be used in conjunction with other purification procedures, such as SEC, to improve the level of purity and yield.

Nevertheless, UF has certain drawbacks. Repeated utilization or handling of intricate biological fluids can result in membrane fouling, which diminishes effectiveness and may require frequent replacement of the membrane [41, 42]. Subjecting exosomes to high pressure or flow rates can induce shear stress, potentially harming them or modifying their functional characteristics [24]. In addition, small particles or molecules that are comparable in size to exosomes may co-concentrate, necessitating supplementary purification procedures to attain a high level of purity [43].

2.6 Precipitation-Based Methods

Precipitation-based techniques for exosome isolation exploit the ability to alter the solubility of exosomes in a solution, thus enabling their separation from other constituents. These approaches utilize substances that can cause the formation of exosomes from a biological fluid such as blood plasma, serum, urine, or cell culture media.

The standard procedure entails introduction of a precipitation solution to the sample containing exosomes and allowing the mixture to incubate for a specific duration, often at a reduced temperature. The precipitation solution typically includes polymers, such as polyethylene glycol (PEG), which have a stronger affinity for water molecules than for exosomes. Consequently, these polymers efficiently displace the exosomes from the fluid, causing them to clump together and settle. After the incubation period, the material is often centrifuged at a lower speed than in ultracentrifugation. The centrifugal force facilitates the sedimentation of the exosomes into a compact mass at the base of the centrifuge tube. The supernatant, with the unwanted components of the sample, can be separated. The exosome pellet can be reconstituted in a buffer and thereafter subjected to additional purification or analysis as required.

Precipitation-based techniques are used due to their simplicity, relatively affordable nature, and lack of need for specialist equipment such as ultracentrifuges [40]. In addition, they can be expanded for high-capacity samples and are often less demanding in terms of labor. Coughlan et al. found that compared to ultracentrifugation, the precipitation method was six times faster and resulted in ~2.5-fold higher concentration of exosomes per mL [44].

A limitation of this approach is its possible lack of selectivity compared to alternative approaches, as it may inadvertently co-precipitate other vesicles or protein aggregates alongside the exosomes [45]. This co-precipitation could compromise the purity of the recovered exosomes.

2.7 Immunoaffinity Capture-Based Isolation

Immunoaffinity capture-based isolation leverages the specificity of antibodies to capture exosomes based on the expression of specific surface markers [21]. Antibodies are selected for their ability to specifically bind to commonly found surface markers on exosomes, such as tetraspanins (CD9, CD63, CD81), or other proteins known to be highly concentrated on the exosomes of interest. The choice of antibodies is crucial and can be customized based on the origin of the exosomes and the objective of isolation. The chosen antibodies are fixed onto a stable substrate, such as magnetic beads, agarose beads, or plates. This immobilization is essential for the subsequent capture and isolation of exosomes from a sample. The exosome-containing biological sample is incubated with antibody-coated beads or plates under conditions that promote the binding of exosomes to the antibodies. During this incubation period, the antibodies effectively and specifically capture exosomes by attaching to their distinct surface biomarkers. Following incubation, the sample is washed to eliminate any unattached components and materials that have bound nonspecifically. This results in the retention of only the exosomes that have been trapped by the antibodies and are firmly attached to the solid support. The captured exosomes are subsequently released into solution from the beads or plate by elution with a buffer that disrupts the binding between the antibodies and the surface proteins of the exosomes. The eluted exosomes can be studied, with characterization of their protein, lipid, and nucleic acid components, as well as their function.

This approach offers a high level of specificity by selectively targeting well-known surface markers found on exosomes, enabling the extraction of specific subsets of exosomes [46]. It tends to produce exosomes with a high level of purity, which is important for subsequent uses such as the identification of biomarkers and therapeutic applications [47, 48]. However, the requirement for precise antibodies can result in higher expenses when compared to alternative exosome isolation procedures. The effectiveness of the isolation process relies on the presence and excellence of antibodies that target the specific exosome indicators of interest. The procedure may be less suitable for high-throughput applications due to the necessary handling and incubation phases.

3 Emerging Techniques: Microfluidic-Based Exosome Isolation

Microfluidic-based exosome isolation techniques involve the use of microscale fluidic devices to separate exosomes from biological materials such as blood, urine, or cell culture media [49, 50]. These strategies leverage the distinctive physical and biological characteristics of exosomes. Microfluidics technology provides numerous benefits compared to conventional exosome isolation methods, such as decreased sample volume, accelerated processing times, enhanced sensitivity, and the possibility of integration with subsequent analysis procedures [50, 51]. Microfluidic devices used for exosome isolation typically utilize one or more of the following principles:

- Size-dependent separation: Microfluidic devices can effectively separate exosomes from other cellular components by exploiting the disparity in size between exosomes (30–150 nm) and larger cellular debris or smaller molecules. This is achieved by the use of filters or nanostructures that are specifically engineered to either trap or permit the passage of particles depending on their size [52, 53].
- Immunoaffinity capture: Microfluidic devices can be modified with antibodies that specifically bind to exosome surface indicators such as CD63, CD9, and CD81, similar to traditional immunoaffinity methods. Exosomes exhibit particular binding affinity to these antibodies when the sample passes through the device, enabling their targeted isolation [54–57].
- Acoustic separation: This technique employs ultrasonic waves to distinguish exosomes from other particles by exploiting disparities in their acoustic characteristics. Exosomes can be partitioned into distinct channels within a microfluidic device without a need for physical filters or chemical markers [58].
- Dielectrophoresis: Dielectrophoresis uses nonuniform electric fields to separate exosomes based on their dielectric characteristics. The particles in a sample experience a force that causes them to migrate toward the electrodes. The magnitude of this force depends on the size and composition of the particles [59, 60].
- Microfluidic inertial focusing: This utilizes the fluid dynamics in small channels to separate and concentrate exosomes. Under precise flow conditions, particles of varying sizes move to distinct positions across the cross-section of the channel, allowing separation based on size [50, 61].

Microfluidic devices have the ability to effectively isolate exosomes with a high level of purity because they give precise control and optimization of conditions. Required sample amounts are usually small, which is especially advantageous when dealing with valuable or scarce materials. Microfluidic devices can be fabricated to combine isolation with subsequent analysis, such as PCR, sequencing, or proteomics, increasing efficiency [62–64]. Microfluidic techniques can greatly decrease processing time in comparison to conventional methods, providing nearly instantaneous processing capabilities. These devices possess the capacity for scalability and automation, which is essential for conducting high-throughput screening and clinical diagnostics. However, the process of creating and producing personalized microfluidic devices can be intricate and expensive, although this is improving with advancements in manufacturing methodologies. Standardization in device design and operation is necessary to guarantee uniformity and replicability of outcomes across many laboratories and research projects. Effective execution may necessitate an interdisciplinary approach encompassing biology, engineering, and materials science.

4 Industrial Methods for Exosome Isolation

In research, selecting an appropriate exosome isolation method for each study depends on various factors such as laboratory conditions, sample volume, sample purity requirements, and budget. Ultracentrifugation is commonly used under experimental conditions to isolate exosomes from cell culture media because it is a simple method that can be stably established in different laboratories and has low cost (apart from the purchase of the machine), high reproducibility, and high yield [15, 44]. However, exosomes obtained from this method may not have sufficiently high purity. In cases demanding high purity, other exosome isolation techniques can be combined.

Unlike research, industrial scale production for therapeutic application may use large volumes of cell culture media, possibly tens of liters, making ultracentrifugation unsuitable. (See Table 2.2 for a summary of scalable techniques.) Recently, TFF has been considered one of the most suitable techniques to harvest exosomes from cell culture media. This technique can collect exosomes from tens of liters of cell culture media [65]. The use of disposable filter columns along with a sanitizable piping system makes this technique a promising candidate for use on an industrial scale [2, 52]. TFF can help harvest exosomes in different sizes based on the pore size of the filter. To simplify TFF, some manufacturers have integrated the filtration process with an automatic control system, creating a fully automated TFF system that facilitates and enhances the efficiency of exosome production (Fig. 2.2). The TFF system can also directly connect to the bioreactor, which helps the isolation process continuously.

5 Conclusion

Exosomes, a type of vesicle secreted by cells, carry important information that can be used therapeutically and diagnostically. For example, exosomes from stem cells may stimulate tissue regeneration. Extracting exosomes from cell culture media is a critical step in their harvesting and production process. The method used determines the yield, recovery efficiency, and purity of the exosome-containing preparation, thereby affecting its safety and effectiveness. Despite much research into various methods for harvesting exosomes from cell culture media, selecting a method for the production of therapeutic products remains a matter of debate. Currently, TFF is a technique with significant potential for separating exosomes from cell culture media to produce products for human application. This technique meets most industrial-scale production requirements while ensuring the quality of the product after separation.

Table 2.2 The scalability of exosome extraction methods

Method	Scalability	Purity	Yield	Cost	Equipment	Industrial applications
Tangential flow filtration (TFF)	High	Moderate	High	Moderate	TFF system	Ideal for large-scale production of exosomes for therapeutics and diagnostics
Size exclusion chromatography (SEC)	High	Moderate	Moderate	Moderate	Chromatography system	Suitable for large-scale purification, often combined with TFF for higher throughput
Polymer-based precipitation	High	Low	High	Low	Centrifuge	Cost-effective for large volumes, but requires additional purification steps for high purity
Continuous flow ultracentrifugation	High	High	High	High	Ultracentrifuge	Used in industrial settings for high-purity exosome isolation, but expensive
Microfluidic technologies	Moderate–high	High	Moderate	High	Microfluidic device	Emerging technology for high-throughput isolation, but not yet widely adopted industrially
Immunoaffinity capture	Moderate–high	High	Low–moderate	High	Beads/columns	Suitable for isolating specific exosome subpopulations, but limited by antibody costs

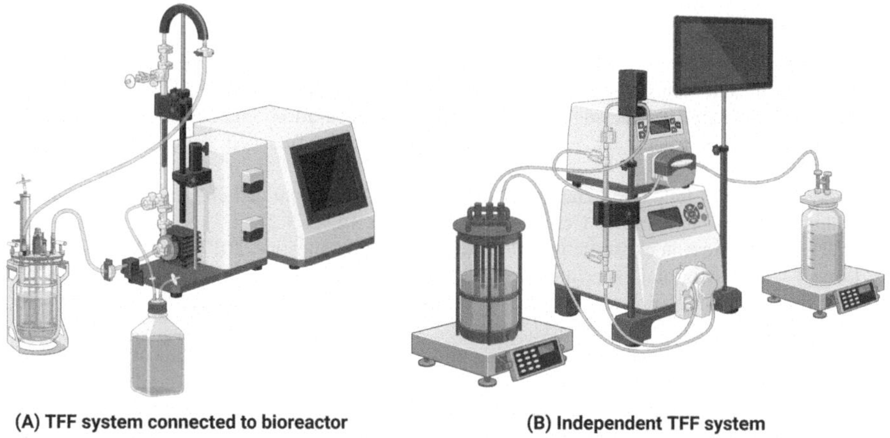

Fig. 2.2 Tangential flow filtration system (TFF). (**a**) TFF system connected to bioreactor: This setup integrates the TFF system directly with a bioreactor, allowing for continuous processing. (**b**) Independent TFF system: This setup operates separately from a bioreactor. The TFF system is standalone, with processing occurring independently of any direct connection to a bioreactor. (The figure is created using biorender.com)

References

1. Chen Y, et al. Protocol for isolating and identifying small extracellular vesicles derived from human umbilical cord mesenchymal stem cells. STAR Protoc. 2024;5(3):103197.
2. Luo H, et al. Large-scale isolation of exosomes derived from NK cells for anti-tumor therapy. Bio Protoc. 2023;13(11):e4693.
3. González-Cubero E, et al. Isolation and characterization of exosomes from adipose tissue-derived mesenchymal stem cells. J Anat. 2021;238(5):1203–17.
4. Zhang M, et al. Methods and technologies for exosome isolation and characterization. Small Methods. 2018;2(9):1800021.
5. Lobb RJ, et al. Optimized exosome isolation protocol for cell culture supernatant and human plasma. J Extracell Vesicles. 2015;4(1):27031.
6. Hassanpour Tamrin S, Sanati Nezhad A, Sen A. Label-free isolation of exosomes using microfluidic technologies. ACS Nano. 2021;15(11):17047–79.
7. Le M-CN, Fan ZH. Exosome isolation using nanostructures and microfluidic devices. Biomed Mater. 2021;16(2):022005.
8. Tayebi M, et al. Exosome purification and analysis using a facile microfluidic hydrodynamic trapping device. Anal Chem. 2020;92(15):10733–42.
9. Livshits MA, et al. Isolation of exosomes by differential centrifugation: theoretical analysis of a commonly used protocol. Sci Rep. 2015;5(1):17319.
10. Helwa I, et al. A comparative study of serum exosome isolation using differential ultracentrifugation and three commercial reagents. PLoS One. 2017;12(1):e0170628.
11. Jeppesen DK, et al. Comparative analysis of discrete exosome fractions obtained by differential centrifugation. J Extracell Vesicles. 2014;3(1):25011.
12. Chen D, et al. Protocol for differential centrifugation-based separation and characterization of apoptotic vesicles derived from human mesenchymal stem cells. STAR Protoc. 2022;3(4):101695.

13. Chhoy P, et al. Protocol for the separation of extracellular vesicles by ultracentrifugation from in vitro cell culture models. STAR Protoc. 2021;2(1):100303.
14. García-Posadas L, et al. Isolation and characterization of human conjunctival mesenchymal stromal cells and their extracellular vesicles. Invest Ophthalmol Vis Sci. 2023;64(12):38.
15. Hu HT, Nishimura T, Suetsugu S. Ultracentrifugal separation, characterization, and functional study of extracellular vesicles derived from serum-free cell culture. STAR Protoc. 2021;2(3):100625.
16. Théry C, et al. Isolation and characterization of exosomes from cell culture supernatants and biological fluids. Curr Protoc Cell Biol. 2006;Chapter 3:Unit 3.22.
17. Li P, et al. Progress in exosome isolation techniques. Theranostics. 2017;7(3):789–804.
18. Lötvall J, et al. Minimal experimental requirements for definition of extracellular vesicles and their functions: a position statement from the International Society for Extracellular Vesicles. J Extracell Vesicles. 2014;3:26913.
19. Xu R, et al. Extracellular vesicle isolation and characterization: toward clinical application. J Clin Invest. 2016;126(4):1152–62.
20. Tauro BJ, et al. Comparison of ultracentrifugation, density gradient separation, and immunoaffinity capture methods for isolating human colon cancer cell line LIM1863-derived exosomes. Methods. 2012;56(2):293–304.
21. Greening DW, et al. A protocol for exosome isolation and characterization: evaluation of ultracentrifugation, density-gradient separation, and immunoaffinity capture methods. Methods Mol Biol. 2015;1295:179–209.
22. Gupta S, et al. An improvised one-step sucrose cushion ultracentrifugation method for exosome isolation from culture supernatants of mesenchymal stem cells. Stem Cell Res Ther. 2018;9(1):180.
23. Van Deun J, et al. The impact of disparate isolation methods for extracellular vesicles on downstream RNA profiling. J Extracell Vesicles. 2014;3. https://doi.org/10.3402/jev.v3.24858
24. Dilsiz N. A comprehensive review on recent advances in exosome isolation and characterization: toward clinical applications. Transl Oncol. 2024;50:102121.
25. Li K, et al. Cushioned-density gradient ultracentrifugation (C-DGUC): a refined and high performance method for the isolation, characterization, and use of exosomes. Methods Mol Biol. 2018;1740:69–83.
26. Brennan K, et al. A comparison of methods for the isolation and separation of extracellular vesicles from protein and lipid particles in human serum. Sci Rep. 2020;10(1):1039.
27. Zhang Z, et al. Comparison of ultracentrifugation and density gradient separation methods for isolating Tca8113 human tongue cancer cell line-derived exosomes. Oncol Lett. 2014;8(4):1701–6.
28. Zhou B, et al. Application of exosomes as liquid biopsy in clinical diagnosis. Signal Transduct Target Ther. 2020;5(1):144.
29. Sidhom K, Obi PO, Saleem A. A review of exosomal isolation methods: is size exclusion chromatography the best option? Int J Mol Sci. 2020;21(18):6466.
30. Mol EA, et al. Higher functionality of extracellular vesicles isolated using size-exclusion chromatography compared to ultracentrifugation. Nanomedicine. 2017;13(6):2061–5.
31. Monguió-Tortajada M, et al. Extracellular vesicle isolation methods: rising impact of size-exclusion chromatography. Cell Mol Life Sci. 2019;76(12):2369–82.
32. Soares M, et al. Isolation of extracellular vesicles from human follicular fluid: size-exclusion chromatography versus ultracentrifugation. Biomol Ther. 2023;13(2):278.
33. Grenhas M, et al. Size-exclusion chromatography: a path to higher yield and reproducibility compared to sucrose cushion ultracentrifugation for extracellular vesicle isolation in multiple myeloma. Int J Mol Sci. 2024;25(15):8496.
34. Doyle LM, Wang MZ. Overview of extracellular vesicles, their origin, composition, purpose, and methods for exosome isolation and analysis. Cells. 2019;8(7):727.
35. Koch LF, et al. Novel insights into the isolation of extracellular vesicles by anion exchange chromatography. Front Bioeng Biotechnol. 2023;11:1298892.

36. Malvicini R, et al. Ion exchange chromatography as a simple and scalable method to isolate biologically active small extracellular vesicles from conditioned media. PLoS One. 2023;18(9):e0291589.
37. Silva RM, et al. Anion exchange chromatography-based platform for the scalable purification of extracellular vesicles derived from human mesenchymal stromal cells. Sep Purif Technol. 2023;310:123238.
38. Zhang W-Y, et al. Large-scale separation and purification of exosomes using ionexchange chromatography. Biomed Res Ther. 2024;11(4):6348–56.
39. Kosanović M, et al. Ion-exchange chromatography purification of extracellular vesicles. BioTechniques. 2017;63(2):65–71.
40. Ansari FJ, et al. Comparison of the efficiency of ultrafiltration, precipitation, and ultracentrifugation methods for exosome isolation. Biochem Biophys Rep. 2024;38:101668.
41. Lebreton B, Brown A, van Reis R. Application of high-performance tangential flow filtration (HPTFF) to the purification of a human pharmaceutical antibody fragment expressed in Escherichia coli. Biotechnol Bioeng. 2008;100(5):964–74.
42. Heinemann ML, et al. Benchtop isolation and characterization of functional exosomes by sequential filtration. J Chromatogr A. 2014;1371:125–35.
43. Gao J, et al. Recent developments in isolating methods for exosomes. Front Bioeng Biotechnol. 2022;10:1100892.
44. Coughlan C, et al. Exosome isolation by ultracentrifugation and precipitation and techniques for downstream analyses. Curr Protoc Cell Biol. 2020;88(1):e110.
45. Yakubovich EI, Polischouk AG, Evtushenko VI. Principles and problems of exosome isolation from biological fluids. Biochem (Mosc) Suppl Ser A Membr Cell Biol. 2022;16(2):115–26.
46. Khanabdali R, et al. High-throughput surface epitope immunoaffinity isolation of extracellular vesicles and downstream analysis. Biol Methods Protoc. 2024;9(1):bpae032.
47. Mondal SK, Whiteside TL. Immunoaffinity-based isolation of melanoma cell-derived and T cell-derived exosomes from plasma of melanoma patients. Methods Mol Biol. 2021;2265:305–21.
48. Sharma P, et al. Immunoaffinity-based isolation of melanoma cell-derived exosomes from plasma of patients with melanoma. J Extracell Vesicles. 2018;7(1):1435138.
49. Wang J, et al. Towards microfluidic-based exosome isolation and detection for tumor therapy. Nano Today. 2021;37:101066.
50. Wu Y, et al. Microfluidic technology for the isolation and analysis of exosomes. Micromachines (Basel). 2022;13(10):1571.
51. Zhao L, et al. Microfluidic-based exosome isolation and highly sensitive aptamer exosome membrane protein detection for lung cancer diagnosis. Biosens Bioelectron. 2022;214:114487.
52. Han Z, et al. Highly efficient exosome purification from human plasma by tangential flow filtration based microfluidic chip. Sensors Actuators B Chem. 2021;333:129563.
53. Shin S, et al. Separation of extracellular nanovesicles and apoptotic bodies from cancer cell culture broth using tunable microfluidic systems. Sci Rep. 2017;7(1):9907.
54. Chen C, et al. Microfluidic isolation and transcriptome analysis of serum microvesicles. Lab Chip. 2010;10(4):505–11.
55. Kanwar SS, et al. Microfluidic device (ExoChip) for on-chip isolation, quantification and characterization of circulating exosomes. Lab Chip. 2014;14(11):1891–900.
56. Im H, et al. Label-free detection and molecular profiling of exosomes with a nano-plasmonic sensor. Nat Biotechnol. 2014;32(5):490–5.
57. Zhang P, He M, Zeng Y. Ultrasensitive microfluidic analysis of circulating exosomes using a nanostructured graphene oxide/polydopamine coating. Lab Chip. 2016;16(16):3033–42.
58. Lee K, et al. Acoustic purification of extracellular microvesicles. ACS Nano. 2015;9(3):2321–7.
59. Lan M, Yang F. Applications of dielectrophoresis in microfluidic-based exosome separation and detection. Chem Eng J. 2024;491:152067.
60. Ayala-Mar S, Gallo-Villanueva RC, González-Valdez J. Dielectrophoretic manipulation of exosomes in a multi-section microfluidic device. Mater Today Proc. 2019;13:332–40.

61. Pouraria H, Foudazi R, Houston JP. Exploitation of elasto-inertial fluid flow for the separation of nano-sized particles: simulating the isolation of extracellular vesicles. Cytometry A. 2023;103(10):786–95.
62. Wang C, et al. Microfluidic biochips for single-cell isolation and single-cell analysis of multiomics and exosomes. Adv Sci (Weinh). 2024;11(28):e2401263.
63. Surappa S, et al. Integrated "lab-on-a-chip" microfluidic systems for isolation, enrichment, and analysis of cancer biomarkers. Lab Chip. 2023;23(13):2942–58.
64. Ramnauth N, et al. Development of a microfluidic device for exosome isolation in point-of-care settings. Sensors. 2023;23(19):8292.
65. Börger V, et al. Tangential flow filtration, a potential method for the scaled preparation of extracellular vesicles. Cytotherapy. 2019;21(5):S57.

Chapter 3
Quality Control of Extracellular Vesicles Derived from Stem Cells for Clinical Applications

Phuc Van Pham

1 Introduction

Extracellular vesicles (EVs), including exosomes, are secreted by stem cells during their growth. These vesicles are rich in proteins, lipids, and nucleic acids, which serve as a means of cell-to-cell communication [1]. The content of these vesicles can vary significantly with changing conditions. In addition, the sensitivity of these information-carrying structures to extracellular factors poses challenges in maintaining the quality of EV products for therapeutic applications [2]. Therefore, ensuring the quality of EV products, including exosomes, is critical in their production for disease treatment. Ineffective quality control not only reduces efficacy but also may cause serious side effects [3]. Three primary reasons for EV quality control are product safety, efficacy, and stability during production.

To ensure safety, EV products must be free from harmful contaminants that could adversely affect patient health. Such contaminants include microorganisms (bacteria, fungi, viruses), endotoxins, mycoplasma, cell debris, and production-introduced contaminants like bovine serum albumin (BSA) [4]. Endotoxins can elicit strong immune responses if present in EV products [5], and cell debris may trigger unwanted effects or immune reactions [6]. Thus, rigorous testing for microorganisms, endotoxins, and other contaminants is essential to minimize risks when EVs are used in humans [7].

For efficacy, EV products must possess high purity and retain biological activity. Assessing EV purity is crucial in the quality control process, as products may be contaminated with protein aggregates or lipoprotein particles [8]. The product's

P. V. Pham (✉)
VNUHCM-US Stem Cell Institute, University of Science, Viet Nam National University Ho Chi Minh City, Ho Chi Minh, Vietnam
e-mail: phucpham@sci.edu.vn

P. V. Pham, S. Fagoonee (eds.), *Stem Cell-Derived Exosomes and Microvesicles*, Stem Cells in Clinical Applications, https://doi.org/10.1007/978-3-031-90698-5_3

purity depends heavily on the technology used for isolating EVs from stem cell culture media. Both scientific and regulatory standards require this purity for clinical applications [2]. The biological activity of EV products is crucial for their effectiveness, their capability to transmit essential information to target cells [9]. Damage or degradation of proteins and nucleic acids within the EVs during production may diminish activity [10]. Evaluation of the biological activity is pivotal for predicting the positive effects of EV products in humans; this involves complex assessments, such as examining immunomodulatory capabilities, angiogenesis stimulation, or tissue regeneration in vitro or in vivo [11].

Ensuring the stability and consistency of product quality between batches is essential for human use and compliance with regulatory standards. Variability in batch quality arises from differences in production conditions or stem cell source quality, complicating treatment protocol consistency [12]. Manufacturers must implement strict production processes and robust quality control methods to monitor EV biological activity, examining aspects like particle size, concentration, and composition [13]. Omics-based techniques offer comprehensive insights into EV composition, helping to ensure batch-to-batch reproducibility [14].

Quality control of EVs is a vital step in their production. By establishing stringent quality control criteria and highly reproducible processes, manufacturers can overcome challenges associated with the quality of EV products. This chapter outlines the key techniques used in EV quality control (Fig. 3.1).

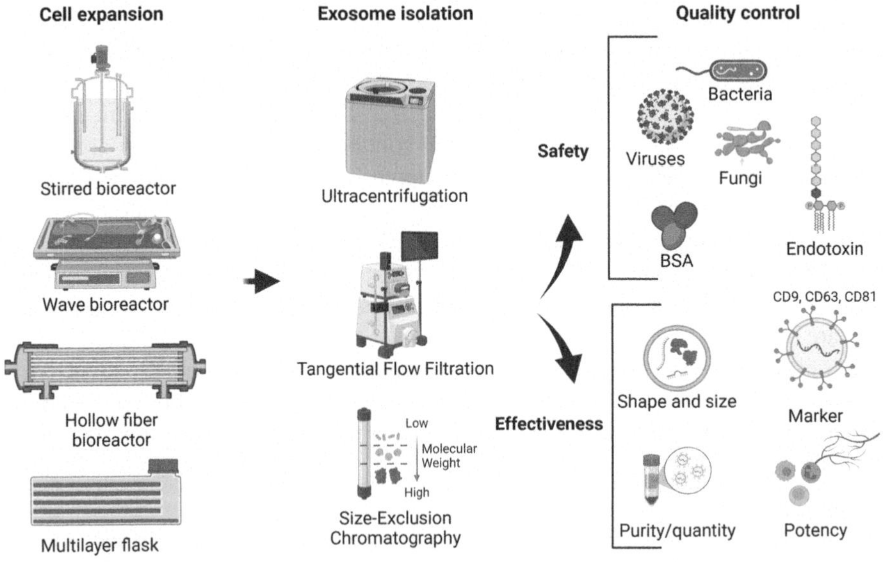

Fig. 3.1 Exosome manufacturing and quality control. Cells are expanded in bioreactors or flasks to obtain cell supernatant. The supernatant is used to isolate exosomes through various methods, including ultracentrifugation, tangential flow filtration, and size-exclusion chromatography. (The figure is created using biorender.com)

2 Quality Control of Stem Cells Derived from Extracellular Vesicles

The quality control of EVs must begin with the stem cells used in the EV production. The quality of stem cells directly affects the quality of EVs. The stem cell health determines the RNA and protein content in the microvesicles (MVs), influencing the EV potency. Contamination of stem cells with mycoplasma or microorganisms can cause corresponding contamination in the EV products. Some critical, controllable attributes of EV products include morphology, size, marker expression, and purity.

2.1 EV Characterization

Employing a combination of measurement methods can yield comprehensive data on the size and characteristics of EV particles. No single technique can fully provide all information about EVs due to their complexity in structure and shape [14].

2.1.1 Morphology

EVs should be nonagglomerated, dish-like, or cup-like structures with clearly defined edges and membranes. Their morphology should be examined using a transmission electron microscope (TEM). Sample preparation requires analytical-grade reagents at room temperature. The EV samples should be placed on a Formvar-carbon copper grid. After 20 min at room temperature, rinse the grid three times with PBS and fix the samples using 1% glutaraldehyde for 5 min. Following fixation, rinse the sample with deionized water eight times. Stain the sample with saturated uranyl oxalate for 5 min before allowing it to dry for 10 min. The sample can then be observed and imaged using the TEM.

2.1.2 Particle Size

EVs derived from stem cells constitute a heterogeneous collection with a range of sizes from 30 nanometers to several micrometers in diameter. Various EV populations within this mixture include exosomes (30–150 nm), microvesicles (100–1000 nm), and apoptotic bodies (500–2000 nm). The isolation process determines the population characteristics, either a mixture of different-sized EV types or a single type. Exosomes in the 30–150 nm size range are the most commonly isolated and utilized. The sizes of exosome particles can be assessed using nano-flow cytometry techniques [15] or nanoparticle tracking analysis [16]. Six methods are available for measuring EV particle size. The advantages and disadvantages of each

Table 3.1 Strengths and limitations of current EV particle size measurement techniques

Method	Strengths	Limitations
Nanoparticle Tracking Analysis (NTA)	High resolution, size distribution, concentration, polydisperse samples	Sensitive to impurities, limited for <50 nm, time-consuming
Dynamic Light Scattering (DLS)	Rapid, non-invasive, minimal preparation	Biased for larger particles, no concentration data, limited for <50 nm
Transmission Electron Microscopy (TEM)	High resolution, morphological details, small particle detection	Extensive preparation, low throughput, artifacts from drying
Tunable Resistive Pulse Sensing (TRPS)	Accurate size and concentration, no labeling required	Requires calibration, sensitive to viscosity, low throughput
Flow cytometry	High-throughput, fluorescent labeling for specific EVs	Limited for small EVs, background noise, expensive
Atomic Force Microscopy (AFM)	3D topography, high resolution, no labeling required	Time-consuming, low throughput, expensive, surface-bound only

method are shown in Table 3.1. Technique selection is usually determined by the study's specific requirements, such as the need for high-resolution, concentration data, or morphological information.

Nanoparticle Tracking Analysis (NTA) NTA utilizes the principles of Brownian motion and light scattering to determine EV size. In this method, a laser beam illuminates particles in a liquid suspension, while a high-sensitivity camera records their movements. The analysis of Brownian motion allows size calculation through the Stokes-Einstein equation, which links particle size to diffusion rate [17]. NTA delivers high-resolution size distribution data for particles ranging from 10 nm to 2000 nm, making it particularly suited for EVs and exosomes. In addition to size data, NTA can estimate particle concentration in the suspension, facilitating quantitative studies [13]. Accurate NTA measurements require meticulous sample preparation to prevent aggregation and contamination by other particles [18].

Dynamic Light Scattering (DLS) Also known as photon correlation spectroscopy, DLS measures fluctuations in scattered light due to the Brownian motion of particles in a solution. These light intensity fluctuations are used to calculate the hydrodynamic diameter, which includes the particle size plus any associated solvent molecules [19]. DLS is a rapid and noninvasive method suitable for monodisperse samples but may be less precise for polydisperse populations due to a bias toward larger particles, which scatter more light. This technique typically measures particles from 1 nm to 10 μm, but its limitations in resolving heterogeneous EV samples make it less ideal for complex EV analysis [20].

Transmission Electron Microscopy (TEM) TEM offers high-resolution imaging for directly observing and accurately sizing EVs. With samples prepared as described above, TEM can image particles as small as 1 nm, providing detailed particle information, including the EVs' bilayer membrane structure. However, the sample preparation and drying process can alter the natural size and shape of EVs [8]. Cryo-TEM, an advanced form of TEM, minimizes these effects by rapid freezing of samples, preserving EVs in a more natural state for accurate size measurement [21].

Tunable Resistive Pulse Sensing (TRPS) TRPS measures particle size by detecting changes in electrical resistance as particles pass through a nanopore, with resistance changes proportional to particle size [22]. TRPS can analyze EVs from 40 to 2000 nm and determine particle concentration, making it useful for characterizing complex EV samples [23]. However, the technique demands careful calibration and optimization of nanopore size and has a relatively low throughput compared to other methods [22].

Flow Cytometry (FCM) FCM is popular for analyzing larger EVs, such as microvesicles, and can extend to smaller EVs by use of fluorescent labels or high-sensitivity instruments [24]. In FCM, EVs tagged with fluorescent markers pass through a laser beam, where particle size and fluorescence are measured. Although capable of high-performance analysis, FCM's ability to analyze particles smaller than 200 nm is limited, as signals may overlap with background noise [25]. Advances in high-resolution FCM have improved sensitivity for smaller EV particles, yet challenges remain in using it for comprehensive EV analysis [26].

Atomic Force Microscopy (AFM) AFM employs a cantilever with a sharp tip to scan the surface of EVs on a substrate, providing information on their shape and allowing accurate 3D size measurement [27]. AFM can resolve particles only a few nanometers in size and is particularly valuable for studying the mechanical properties of EVs [28]. However, AFM is time-intensive and requires careful sample preparation to prevent inaccuracies caused by EV shape alterations from the tip [27].

2.1.3 Marker Expression

The use of markers in exosomes allows the confirmation of their origin, composition, and function. Common surface protein markers of exosomes utilized for identification include tetraspanins (such as CD9, CD63, and CD81), which play crucial roles in vesicle formation and cargo sorting [4, 29]. Exosomes are also distinguished by markers associated with the endosomal sorting complex required for transport (ESCRT), including Alix and TSG101 [25, 30]. Heat shock proteins (HSP70, HSP90) and flotillin-1 can also be used to confirm the presence of exosomes [11, 31]. In addition to the presence of specific markers, the absence of proteins like calnexin (an endoplasmic reticulum marker) and cytochrome c (mitochondrial marker) is used to assess the purity of exosomes [2, 8] (Table 3.2).

Table 3.2 Differences in markers between exosomes, microvesicles, and apoptotic bodies

Feature	Exosomes	Microvesicles (MVs)	Apoptotic bodies (ABs)
Size	30–150 nm	100–1000 nm	500–2000 nm (larger)
Origin	Endosomal pathway (multivesicular bodies)	Plasma membrane budding	Cell disintegration during apoptosis
Markers	Tetraspanins (CD9, CD63, CD81)	Integrins (CD41, CD61)	Caspases
	ESCRT proteins (TSG101, Alix)	Selectins (CD62P)	Histones (e.g., H2B)
	Heat shock proteins (HSP70, HSP90)	Phosphatidylserine (Annexin V)	DNA fragments
	Endosomal markers (e.g., Rab GTPases)	ARF6, VCAMP3	Phosphatidylserine (Annexin V)
Unique features	Enriched in endosomal proteins	Directly bud from plasma membrane	Contain nuclear and cytoplasmic debris
	Carry mRNA, miRNA, and proteins	Heterogeneous in size and content	Larger size and irregular shape
Detection methods	Flow cytometry	Flow cytometry	Flow cytometry
	Electron microscopy	Electron microscopy	Electron microscopy
	Proteomic analysis	Proteomic analysis	Proteomic analysis

Beyond exosomes, microvesicles (MVs) and apoptotic bodies (ApoBDs) can be identified by the expression of specific markers. Since MVs often form directly from the plasma membrane, they typically display integrins (such as CD41, CD61) and selectins (such as CD62P) as well as phosphatidylserine (detected by Annexin V) on their surface [4, 6]. MVs also express markers like ARF6 and VCAMP3 related to their biosynthesis [32, 33]. ApoBDs are larger than MVs and are formed as a result of programmed cell death (apoptosis); hence, they often express caspases, histones, DNA fragments, and some specific markers of their original source cells [34, 35]. ApoBDs also express phosphatidylserine on the outer membrane surface (as a result of apoptosis), but they differ from MVs in their larger size compared to MVs and in the concurrent presence of nuclear components (such as histone H2B) [36, 37].

Various techniques are available for marker detection: FCM, Western blotting, enzyme-linked immunosorbent assay (ELISA), TEM, NTA, mass spectrometry (MS), and RNA sequencing and PCR (Table 3.3).

FCM can be effective in evaluating EV markers as they pass through a laser beam while suspended in solution. The scattered light and fluorescence emitted from the EV particles, when labeled with fluorescent-tagged markers, can be analyzed to identify specific markers. FCM enables the simultaneous detection of multiple markers. For example, the use of monoclonal antibodies labeled with fluorescent dyes specific for CD9, CD63, and CD81 can aid in exosome identification. However, because the small size of exosomes can lead to confusion with background noise, some advanced FCM techniques employ large latex beads (ranging

Table 3.3 Advantages and disadvantages of some techniques for size determination of EVs

Technique	Advantages	Disadvantages
Flow cytometry	High-throughput analysis Detects multiple markers simultaneously Provides size and concentration data	Limited resolution for small EVs (<100 nm) Requires expensive equipment and expertise
Western blotting	Highly specific for protein detection Validates multiple markers Widely accessible	Semi-quantitative Time-consuming Requires large sample volumes
ELISA	High sensitivity and specificity Suitable for high-throughput screening Quantifies markers in complex fluids	Limited to known markers Requires specific antibodies Expensive for multiplexing
Transmission electron microscopy (TEM)	High-resolution imaging of EV morphology Can confirm specific markers with immunogold labeling	Expensive and labor-intensive Requires specialized training Low throughput
Nanoparticle tracking analysis (NTA)	Measures size distribution and concentration Can combine with fluorescent labeling Analyzes EVs in native state	Limited to size and concentration Less effective for polydisperse samples Expensive equipment
Mass spectrometry (MS)	Highly sensitive and specific Provides comprehensive protein/lipid profiles Identifies novel biomarkers	Expensive and complex Requires extensive sample preparation\data analysis can be challenging
RNA sequencing and PCR	Identifies RNA biomarkers High sensitivity and specificity Provides functional insights	Expensive for RNA-seq Requires high-quality RNA Complex data analysis

from 5 to 10 micrometers) that capture EVs on their surfaces before being stained with monoclonal antibodies. High-resolution flow cytometry can directly analyze EV particles without the need for latex bead carriers.

Western blotting allows the separation of EV proteins by their molecular weight through gel electrophoresis. In this method, the total protein from EVs is separated and transferred onto a membrane, followed by staining with antibodies linked to probes that detect protein markers (e.g., CD9, CD63, TSG101). The signals from these probes are read using chemiluminescence or fluorescence, facilitating the identification of specific markers.

ELISA enables the detection and quantification of protein markers in EV samples. In this process, total proteins from EVs are captured by specific antibodies on the plate surface, followed by a secondary antibody to detect the presence of protein markers. Typically, secondary antibodies are linked to an enzyme or fluorescent molecule to measure the concentration of target proteins. This technique is highly sensitive and specific, suitable for large-scale screening [38].

TEM is used to visualize the morphology and size of EV particles. When EV particles are stained with monoclonal antibodies specific to markers and tagged with gold particles (immunogold labeling), TEM can be used to evaluate the expression of surface markers on EVs [39].

Like TEM, NTA can be used to detect surface markers of EVs when markers are labeled with fluorescent-tagged antibodies [40].

MS can identify the presence of EV markers after lysis of EVs and ionization of their components, which are then separated using a mass spectrometer. Specific markers, such as tetraspanins and heat shock proteins, can be identified from the obtained spectra [41]. MS also can be used to discover new EV markers [42].

Other widely used techniques for recognizing the presence of EV markers and components include RNA sequencing and PCR. These methods frequently identify RNA markers within EVs and demonstrate the presence of microRNAs [43].

Each technique for assessing EV marker presence has advantages and drawbacks. FCM and NTA are ideal for analyzing particle size and concentration, while Western blotting, ELISA, and TEM are generally used to provide insights into protein markers. MS and RNA sequencing offer information about the molecular composition of nucleic acids inside the EVs. Combining these techniques yields comprehensive information on the contents and markers of EVs.

2.2 *The Purity of EVs*

The purity of EVs influences their efficacy and adverse effects. EV products may be contaminated with other cellular debris or proteins from the culture medium. Therefore, the purity of EVs should be assessed for the presence of other particles and protein contamination [3, 4].

A combination of techniques should be employed to evaluate particle purity and contamination from other particles. For example, to assess the purity of exosome products (typically 30–150 nm), methods such as NTA or DLS can determine the size distribution and concentration of exosomes in the sample [44, 45]. Western blotting or FCM techniques can detect specific exosome markers like CD9, CD63, CD81, and TSG101 and ensure the absence of markers of contaminating particles (such as microvesicles) that express apolipoprotein or endoplasmic reticulum markers [4, 9]. TEM is then employed to observe the shapes of the exosomes [46]. Finally, techniques such as MS or RNA profiling should be used to assess exosome purity by analyzing specific proteins and nucleic acids within the exosomes [4, 8].

In addition to assessing contamination by other vesicles, contamination from proteins in the culture medium, such as albumin and serum proteins, must also be evaluated in the EV product. These proteins can typically be detected by methods like SDS-PAGE (sodium dodecyl sulfate-polyacrylamide gel electrophoresis), Western blotting, or ELISA [3, 4]. In some instances, MS can also be employed to detect foreign proteins in the product [2, 8].

2.3 *Safety Parameters*

2.3.1 Microorganism Contamination

Ensuring the control of microbial contamination in EV products is crucial for the safety and efficacy of these products, especially when they are intended for therapeutic or diagnostic purposes. Rigorous testing is required to verify that the product is free from bacteria, fungi, viruses, and mycoplasma [4]. These tests must adhere to the guidelines and regulations governing biopharmaceuticals, such as those outlined by the International Council for Harmonisation (ICH) and the US Pharmacopeia (USP) (FDA, 2020; USP <71>, 2023). To confirm that the samples meet microbial standards, it is essential to carefully screen stem cell donors, and the stem cell sources utilized in the manufacturing of EV products must be evaluated for sterility in advance. In general, assessments should be carried out from the raw materials stage through the entire process to the final product [47].

For example, when producing exosome products from adipose-derived stem cells that meet sterility requirements, it is imperative to thoroughly screen the fat donors for infections such as HIV 1/2, HBV, HCV, CMV, HTLV, and syphilis, as recommended by the American Association of Tissue Banks [48]. Sterility evaluation after isolation of the stem cells must ensure a lack of contamination, following protocols such as those by USP 71. Only when the stem cell source is confirmed to be free from microbial contamination can it be used for exosome production. In addition, the exosome product itself must be monitored for infections and mycoplasma contamination using validated methods like PCR or nucleic acid testing (European Pharmacopoeia 2.6.7, 2023).

2.3.2 BSA Contamination

EVs, including exosomes, are frequently contaminated with BSA from cell culture media components such as fetal bovine serum (FBS). Conventional isolation methods like ultracentrifugation often fail to eliminate BSA due to overlapping physical properties (e.g., size and density) between BSA aggregates and EVs [4, 45]. Residual BSA in exosome preparations poses risks of unintended immune responses in therapeutic applications. To mitigate this, transitioning to BSA-free or serum-free culture systems is recommended, though advanced purification techniques such as size-exclusion chromatography (SEC) [49] or tangential flow filtration (TFF) [50] can also reduce BSA contamination significantly.

Post-isolation quality control via Western blotting or ELISA is critical for detecting residual BSA [3]. However, optimizing exosome yield while maintaining BSA-free conditions remains a challenge, necessitating trade-offs between purity and productivity. Ultimately, the production workflow and permissible BSA thresholds must align with the intended application of the EVs.

2.3.3 Endotoxin

Endotoxins are lipopolysaccharide (LPS) molecules originating from the outer membrane of gram-negative bacteria and are potent inducers of inflammation and sepsis [51]. The various centrifugation and filtration methods used to obtain EV preparations can only partially remove endotoxins [4], which have a strong affinity for EV membranes or may co-precipitate during EV isolation. The absence of standardized procedures for endotoxin detection and removal increases variability among studies on EV products. Recent research demonstrates that using endotoxin-free culture media, avoiding bovine serum, employing advanced purification techniques, and rigorously testing for endotoxins with the limulus amebocyte lysate (LAL) assay (US FDA, 2012) can ensure that EV products meet safety standards for endotoxin levels [52]. Maintaining low endotoxin levels is crucial for the clinical application of exosome products.

For stem cell-derived exosome products used in clinical applications, endotoxin levels must comply with strict standards set by regulatory agencies such as the FDA in the United States (US FDA, 2012) or the EMA in Europe (EMA, 2008). Generally, the maximum allowable limit is $\leq$0.5 endotoxin units (EU)/mL for injectable products or $\leq$ 5.0 EU/kg/hour for intravenous therapies.

2.4 Biological Activity

The biological activity of the EV preparation is a critical criterion for ensuring its therapeutic efficacy in human applications. Activity assessments are typically conducted through in vitro or in vivo evaluations (using animal models). In vivo studies often face challenges related to complexity, ethical concerns, and variability across experimental repetitions, limiting their reproducibility and translational relevance. Consequently, in vitro assays are widely preferred for initial screening due to their scalability, cost-effectiveness, and ethical acceptability. Common in vitro tests to evaluate EV bioactivity include the following:

1. Cell proliferation: Measured using assays such as MTT or BrdU incorporation to quantify EV-induced growth stimulation or inhibition in target cells [3]
2. Cell migration: Evaluated via scratch/wound-healing assays or transwell chambers to assess EVs' role in enhancing or inhibiting cellular motility [4]
3. Differentiation effects: Demonstrated by lineage-specific markers (e.g., osteogenic, adipogenic) in stem cells treated with EVs [5]
4. Anti-inflammatory effects: Analyzed through cytokine profiling (e.g., reduced TNF-α, IL-6) or macrophage polarization assays [6]
5. Angiogenesis effects: Assessed using endothelial tube formation assays (e.g., with HUVECs) to measure pro- or anti-angiogenic potential [7] (Table 3.4).

However, almost all in vitro and in vivo assays for biological activity are complex and costly. Some manufacturers have established tests to control the composition of exosomes, such as examining miRNAs or proteins related to the varying potency of exosomes.

Table 3.4 Some methods used to assess the biological activity of EVs

Function	Method	Purpose	Key steps/readouts	Considerations
Proliferation	Cell/nuclei counting	Assess effects of EVs on target cell proliferation	Seed cells on plates → counting cells/nuclei by hematometter or nuclei counter	Control for proliferation (e.g., mitomycin-C); serum-free conditions
	MTT, XTT, WST-1 or Alarma blue	Assess effects of EVs on target cell proliferation	Seed cells in multiwell plates → measure cell proliferation via metabolic activity using a plate reader	Control for proliferation (e.g., mitomycin-C); serum-free conditions.
Migration	Scratch assay	Assess EV-induced cell migration and wound closure	Create scratch in cell monolayer → treat with EVs → image closure over time → quantify with ImageJ	Control for proliferation (e.g., mitomycin-C); serum-free conditions
	Transwell migration	Quantify directional migration through a membrane	Seed cells in upper chamber → EVs in lower chamber → stain/count migrated cells	Use Matrigel for invasion assays; normalize to baseline migration
Differentiation	Osteogenesis	Test EV-driven bone differentiation (e.g., MSCs → osteoblasts)	Culture MSCs in osteogenic medium + EVs → stain calcium (alizarin red) → measure ALP/Runx2	Long culture period (14–21 days); use differentiation media controls
	Adipogenesis	Test EV-driven fat differentiation (eSCs → adipocytes)	Culture MSCs in adipogenic medium + EVs → stain lipids (oil red O) → measure PPARγ/FABP4.	Synergy with differentiation factors (e.g., dexamethasone)
Anti-inflammatory	Cytokine ELISA	Measure EV-mediated suppression of pro-inflammatory cytokines	Stimulate macrophages (LPS/IFN-γ) → treat with EVs → quantify IL-6/TNF-α via ELISA	Test EVs for endotoxins; use multiplex assays (e.g., Luminex) for multiple targets.
Angiogenesis	Tube formation assay	Evaluate EV-induced blood vessel formation	Seed endothelial cells (HUVECs) on Matrigel + EVs → image tube networks → quantify nodes/junctions	Matrigel batch variability; include VEGF controls
Potency	Dose-response curves	Quantify EV bioactivity strength (e.g., EC50/IC50)	Treat cells with EV dilutions → measure response (proliferation, cytokines, etc.) → fit sigmoidal curve	Normalize EV dose (particles/protein); test batch consistency

3 Conclusion

At each stage of production, quality control of EVs derived from stem cells for clinical applications is crucial to ensuring their safety, efficacy, and reliability as a therapeutic treatment. EV therapies encounter significant challenges due to the variability in EV quality, particularly their sensitivity to production conditions and the high risk of contamination during manufacturing. Rigorous quality control processes must adhere to three key pillars—safety, efficacy, and consistency—across production batches. Safety is paramount to prevent adverse effects in patients. EV products must be meticulously screened for contaminants such as bacteria, viruses, mycoplasma, and endotoxins. Careful donor screening of stem cells, using endotoxin-free culture media, and advanced EV purification techniques are essential to ensure the safety of EV products. The residual presence of BSA in the culture media should be monitored using ELISA or Western blotting techniques. To ensure the efficacy of EV products, both their purity and biological activity need to be evaluated. Purity should be assessed using a combination of techniques including NTA, TEM, MS, and the expression of specific EV markers through FCM; their biological activity can be evaluated through in vitro assessments such as proliferation, migration, and angiogenesis assays on target cells. Achieving consistency across production batches is a significant challenge for the clinical application of EVs. Standardizing processes to enhance reproducibility is essential, as is integrating quality control throughout the entire production process. In conclusion, the clinical application of EVs requires quality control strategies that employ advanced analytical techniques to ensure that the production process conforms to global EV standards.

References

1. Zhang Y, et al. Exosomes: biogenesis, biologic function and clinical potential. Cell Biosci. 2019;9:19.
2. Witwer KW, Théry C. Extracellular vesicles or exosomes? On primacy, precision, and popularity influencing a choice of nomenclature. J Extracell Vesicles. 2019;8(1):1648167.
3. Lötvall J, et al. Minimal experimental requirements for definition of extracellular vesicles and their functions: a position statement from the International Society for Extracellular Vesicles. J Extracell Vesicles. 2014;3:26913.
4. Théry C, et al. Minimal information for studies of extracellular vesicles 2018 (MISEV2018): a position statement of the International Society for Extracellular Vesicles and update of the MISEV2014 guidelines. J Extracell Vesicles. 2018;7(1):1535750.
5. Görgens A, et al. Optimisation of imaging flow cytometry for the analysis of single extracellular vesicles by using fluorescence-tagged vesicles as biological reference material. J Extracell Vesicles. 2019;8(1):1587567.
6. van der Pol E, et al. Classification, functions, and clinical relevance of extracellular vesicles. Pharmacol Rev. 2012;64(3):676–705.
7. Mateescu B, et al. Obstacles and opportunities in the functional analysis of extracellular vesicle RNA – an ISEV position paper. J Extracell Vesicles. 2017;6(1):1286095.

8. Tkach M, Théry C. Communication by extracellular vesicles: where we are and where we need to go. Cell. 2016;164(6):1226–32.
9. Raposo G, Stoorvogel W. Extracellular vesicles: exosomes, microvesicles, and friends. J Cell Biol. 2013;200(4):373–83.
10. Colombo M, Raposo G, Théry C. Biogenesis, secretion, and intercellular interactions of exosomes and other extracellular vesicles. Annu Rev Cell Dev Biol. 2014;30:255–89.
11. Kowal J, et al. Proteomic comparison defines novel markers to characterize heterogeneous populations of extracellular vesicle subtypes. Proc Natl Acad Sci USA. 2016;113(8):E968–77.
12. Geeurickx E, et al. The generation and use of recombinant extracellular vesicles as biological reference material. Nat Commun. 2019;10(1):3288.
13. Gardiner C, et al. Techniques used for the isolation and characterization of extracellular vesicles: results of a worldwide survey. J Extracell Vesicles. 2016;5:32945.
14. Welsh JA, et al. MIFlowCyt-EV: a framework for standardized reporting of extracellular vesicle flow cytometry experiments. J Extracell Vesicles. 2020;9(1):1713526.
15. Tian Y, et al. Quality and efficiency assessment of six extracellular vesicle isolation methods by nano-flow cytometry. J Extracell Vesicles. 2020;9(1):1697028.
16. Comfort N, et al. Nanoparticle tracking analysis for the quantification and size determination of extracellular vesicles. J Vis Exp. 2021;169
17. Filipe V, Hawe A, Jiskoot W. Critical evaluation of nanoparticle tracking analysis (NTA) by NanoSight for the measurement of nanoparticles and protein aggregates. Pharm Res. 2010;27(5):796–810.
18. Dragovic RA, et al. Sizing and phenotyping of cellular vesicles using nanoparticle tracking analysis. Nanomedicine. 2011;7(6):780–8.
19. Pecora R. Dynamic light scattering measurement of nanometer particles in liquids. J Nanopart Res. 2000;2:123–31.
20. Sitar S, et al. Size characterization and quantification of exosomes by asymmetrical-flow field-flow fractionation. Anal Chem. 2015;87(18):9225–33.
21. Wu Y, Deng W, Klinke DJ II. Exosomes: improved methods to characterize their morphology, RNA content, and surface protein biomarkers. Analyst. 2015;140(19):6631–42.
22. Vogel R, et al. A standardized method to determine the concentration of extracellular vesicles using tunable resistive pulse sensing. J Extracell Vesicles. 2016;5(1):31242.
23. Anderson W, et al. A comparative study of submicron particle sizing platforms: accuracy, precision and resolution analysis of polydisperse particle size distributions. J Colloid Interface Sci. 2013;405:322–30.
24. Lacroix R, et al. Overcoming limitations of microparticle measurement by flow cytometry. In: Seminars in thrombosis and hemostasis. © Thieme Medical Publishers; 2010.
25. Van Der Vlist EJ, et al. Fluorescent labeling of nano-sized vesicles released by cells and subsequent quantitative and qualitative analysis by high-resolution flow cytometry. Nat Protoc. 2012;7(7):1311–26.
26. Pospichalova V, et al. Simplified protocol for flow cytometry analysis of fluorescently labeled exosomes and microvesicles using dedicated flow cytometer. J Extracell Vesicles. 2015;4(1):25530.
27. Sharma S, et al. Structural-mechanical characterization of nanoparticle exosomes in human saliva, using correlative AFM, FESEM, and force spectroscopy. ACS Nano. 2010;4(4):1921–6.
28. Yuana Y, et al. Atomic force microscopy: a novel approach to the detection of nanosized blood microparticles. J Thromb Haemost. 2010;8(2):315–23.
29. Zhang H, et al. Identification of distinct nanoparticles and subsets of extracellular vesicles by asymmetric flow field-flow fractionation. Nat Cell Biol. 2018;20(3):332–43.
30. Kalluri R, LeBleu VS. The biology, function, and biomedical applications of exosomes. Science. 2020;367(6478):eaau6977.
31. Mathieu M, et al. Specificities of secretion and uptake of exosomes and other extracellular vesicles for cell-to-cell communication. Nat Cell Biol. 2019;21(1):9–17.

32. Muralidharan-Chari V, et al. ARF6-regulated shedding of tumor cell-derived plasma membrane microvesicles. Curr Biol. 2009;19(22):1875–85.
33. Hsu C, et al. Regulation of exosome secretion by Rab35 and its GTPase-activating proteins TBC1D10A–C. J Cell Biol. 2010;189(2):223–32.
34. Kerr JF, Wyllie AH, Currie AR. Apoptosis: a basic biological phenomenon with wideranging implications in tissue kinetics. Br J Cancer. 1972;26(4):239–57.
35. Elmore S. Apoptosis: a review of programmed cell death. Toxicol Pathol. 2007;35(4):495–516.
36. Poon IK, et al. Apoptotic cell clearance: basic biology and therapeutic potential. Nat Rev Immunol. 2014;14(3):166–80.
37. Caruso S, Poon IK. Apoptotic cell-derived extracellular vesicles: more than just debris. Front Immunol. 2018;9:1486.
38. Kurian TK, et al. Elucidating methods for isolation and quantification of exosomes: a review. Mol Biotechnol. 2021;63(4):249–66.
39. Su Y, Kumar A, Deep G. Characterization of exosomal surface proteins by Immunogold labeling. Methods Mol Biol. 2022;2413:177–82.
40. Desgeorges A, et al. Differential fluorescence nanoparticle tracking analysis for enumeration of the extracellular vesicle content in mixed particulate solutions. Methods. 2020;177:67–73.
41. Newman LA, Useckaite Z, Rowland A. Addressing MISEV guidance using targeted LC-MS/MS: a method for the detection and quantification of extracellular vesicle-enriched and contaminant protein markers from blood. J Extracell Biol. 2022;1(9):e56.
42. Misawa T, et al. Identification of novel senescent markers in small extracellular vesicles. Int J Mol Sci. 2023;24(3)
43. Zhou L, et al. Mesenchymal stem cell-derived exosomes ameliorate TGF-β1-induced endometrial fibrosis by altering their miRNA profile. Am J Transl Res. 2023;15(5):3203–16.
44. Coumans FA, et al. Methodological guidelines to study extracellular vesicles. Circ Res. 2017;120(10):1632–48.
45. Théry C, et al. Isolation and characterization of exosomes from cell culture supernatants and biological fluids. Curr Protoc Cell Biol. 2006;30(1):3–22.
46. Van der Pol E, et al. Recent developments in the nomenclature, presence, isolation, detection and clinical impact of extracellular vesicles. J Thromb Haemost. 2016;14(1):48–56.
47. Witwer KW, et al. Updating MISEV: evolving the minimal requirements for studies of extracellular vesicles. Wiley Online Library; 2021. p. e12182.
48. Banks AAOT. Standards for tissue banking. 5th ed. The Association; 2021.
49. Böing AN, et al. Single-step isolation of extracellular vesicles by size-exclusion chromatography. J Extracell Vesicles. 2014;3(1):23430.
50. Dilsiz N. A comprehensive review on recent advances in exosome isolation and characterization: toward clinical applications. Transl Oncol. 2024;50:102121.
51. Caroff M, Karibian D. Structure of bacterial lipopolysaccharides. Carbohydr Res. 2003;338(23):2431–47.
52. Pan W, et al. Challenges and strategies: scalable and efficient production of mesenchymal stem cells-derived exosomes for cell-free therapy. Life Sci. 2023;319:121524.

Chapter 4
Preservation of Extracellular Vesicles

Nhat Chau Truong, Nghia Minh Do, and Phuc Van Pham

1 Introduction

Exosomes are vesicles secreted from cells with diameters ranging from 30 to 150 nm. These vesicles carry biologically meaningful information. Recent studies have shown that exosomes may contain miRNA molecules, proteins, and even DNA, molecules that encode information for intercellular communication; for example, regulating tissue homeostasis and tissue regeneration after injury [1–3]. Therefore, exosomes, especially those derived from stem cells, have become a research focus due to their potential applications in treating conditions related to disrupted tissue homeostasis and stimulating tissue regeneration [4, 5].

Currently, exosomes have been successfully isolated from various types of stem cells, including mesenchymal stem cells [6, 7], induced pluripotent stem cells [8], endothelial progenitor cells [9, 10], and neural stem cells [11, 12]. Many animal studies have shown that these exosomes are effective in treating various pathological conditions. In 2021, Zhang et al. demonstrated that exosomes from mesenchymal stem cells possessed immunomodulatory properties and reduced inflammation induced by the drug Imiquimod in a psoriasis-like animal model [13]. Exosomes from mesenchymal stem cells from adipose tissue, the amniotic membrane, or the human umbilical cord can improve ovarian function and prevent premature ovarian aging in mice [6, 7, 14]. Mesenchymal stem cell-derived exosomes can treat spinal cord injuries, resulting in reduced inflammation and neural scarring [15–17]. Besides mesenchymal stem cells, exosomes from induced pluripotent stem cells are also employed in clinical treatment. These exosomes have applications in treating

N. C. Truong · N. M. Do · P. V. Pham (✉)
VNUHCM-US Stem Cell Institute, University of Science, Viet Nam National University Ho Chi Minh City, Ho Chi Minh, Vietnam
e-mail: nhattruong@sci.edu.vn; nghiado@sci.edu.vn; phucpham@sci.edu.vn

P. V. Pham, S. Fagoonee (eds.), *Stem Cell-Derived Exosomes and Microvesicles*, Stem Cells in Clinical Applications, https://doi.org/10.1007/978-3-031-90698-5_4

infertility [18], avascular necrosis of the femoral head [19], and reducing vascular remodeling in pulmonary arterial hypertension [8].

Exosomes from adult mesenchymal stem cells have been successfully used in humans. A study by Park et al. (2023) used a combination of exosomes from adipose-derived mesenchymal stem cells and microneedling technology in facial skin rejuvenation. The results showed significant improvement in skin metrics, especially in reducing wrinkles, increasing skin elasticity, and improving pigmentation after 12 weeks compared to the control group [20]. Research by Wang et al. (2023) also indicated that using exosomes from umbilical cord mesenchymal stem cells significantly improved melasma symptoms [21]. Exosomes from adipose-derived mesenchymal stem cells have been tested in treatments for Alzheimer's disease [22].

With such significant potential, the production of exosomes in general and stem cell-derived exosomes in particular has become a heavily invested industry. In addition to research on the refining and isolation processes, the preservation of exosomes is crucial for clinical application. At present, several strategies are employed for exosome preservation, each with its advantages and limitations. Three main preservation methods are currently in use: freezing, lyophilization, and spray drying. However, unlike cryopreserved stem cells [23], which are as effective as fresh stem cells, cryopreserved exosomes are less effective than fresh exosomes [24]. Under freezing conditions, exosomes may undergo changes in RNA composition, altered biological activity, changes in size, and aggregation [25]. However, Susa et al. (2023) showed that if suitable preservatives are used in lyophilization, the rehydrated exosomes can retain their biological activity [26]. Exosomes from mesenchymal stem cells frozen in an environment supplemented with trehalose were able to maintain their shape and function [27]. Hematopoietic stem cells co-cultured with exosomes exosomes preserved both shape and function when frozen [27]. Lo and colleagues (2024) evaluated the impact of lyophilized exosomes on the proliferation and characteristics of ligament cells obtained from rabbits [28]. The reconstituted exosomes significantly enhanced the viability, proliferation, and migration of ligament cells compared to the control group [28]. Furthermore, the levels of collagen, TGF-beta, and VEGF were substantially increased in the exosome-treated group after 24 h [28].

This chapter summarizes and introduces common preservation techniques for exosomes and analyzes the advantages and limitations of each technique (Table 4.1).

2 The Principles of Extracellular Vesicle Preservation

Different from stem cells that have therapeutic effects based on both biomolecules and the participation of the cells in injury healing, the bioeffects of EVs are based on the biomolecules inside vesicles as well as several molecules on the surfaces of the EVs. Thus, EV preservation methods must maintain the structures of proteins, peptides, and acid nucleic acids (DNA, RNA) similar to protein and DNA

Table 4.1 Advantages and disadvantages of some methods used for exosome preservation

Method	Principle	Advantages	Disadvantages	Medium
Cryopreservation	Exosomes are stored at ultra-low temperatures (typically −80 °C or −196 °C) to slow down metabolic and chemical degradation processes	Maintains structural and functional integrity of exosomes Long-term storage capability Widely used and well studied	Requires specialized equipment (e.g., ultra-low freezers, liquid nitrogen) Risk of ice crystal formation, which can damage exosomes High energy consumption	Cryoprotectants (e.g., dimethyl sulfoxide (DMSO), glycerol, or sucrose) are often added to prevent ice crystal formation Exosomes are typically suspended in phosphate-buffered saline (PBS) or cell culture media
Freeze drying	Exosomes are frozen and then subjected to a vacuum to remove water by sublimation, resulting in a dry powder	Extends shelf life at room temperature Reduces storage and transportation costs Minimal risk of contamination	May cause structural damage to exosomes due to ice crystal formation or dehydration stress Requires reconstitution before use, which may affect functionality Time-consuming process	Lyoprotectants (e.g., trehalose, sucrose, or mannitol) are commonly added to protect exosomes during the freeze drying process Exosomes are typically suspended in a buffer (e.g., PBS) before freeze drying
Spray drying	Exosomes are dispersed in a solution and sprayed into a hot gas stream, rapidly drying them into a powder	Fast and scalable process Produces a dry, stable powder suitable for storage at room temperature Cost-effective for large-scale production	High temperatures during drying may damage exosome integrity Risk of aggregation or loss of functionality Requires optimization to minimize heat-induced damage	Protectants (e.g., trehalose, lactose, or mannitol) are often added to the exosome suspension to stabilize the exosomes during the drying process Exosomes are typically suspended in a liquid medium (e.g., PBS or water) before spray drying

preservation. However, the biological effects of EVs also depend on the integrity of the exosome structure. This means that EVs should be “alive” after preservation. The EVs are structured with a double lipoprotein membrane, similar to cells, that protects the molecules inside the vesicles. Under suitable conditions, this membrane can be broken, and the contents of EVs can be lost along with factors outside the vesicles.

There are two main reasons for stipulating that during preservation, EVs should maintain their structure and activity. First, different from single molecules of DNA, RNA, or protein, the information contained in EVs depends on a cocktail of molecules transported from the provider cells to the recipient cells. The composition of types and ratios of biomolecules within EVs creates specialized signals to the recipient cells. Therefore, any changes in the composition or ratios of molecules can affect the signals. Second, in therapeutic applications, if the membranes of EVs are broken, the molecules will be digested by enzymes in the plasma or tissue fluid before they can be transported and delivered to the target cells.

There are two main approaches used to preserve EVs: cryopreservation and lyophilization (Fig. 4.1). In cryopreservation, to avoid the EV membranes from breaking, the water inside EVs is removed before the temperature is lowered, as water can form ice crystals and damage the membrane. In general, a low temperature in combination with dehydration is used to protect the biomolecules inside EVs from enzymatic digestion. Theoretically, EVs can be refrigerated to inhibit enzyme

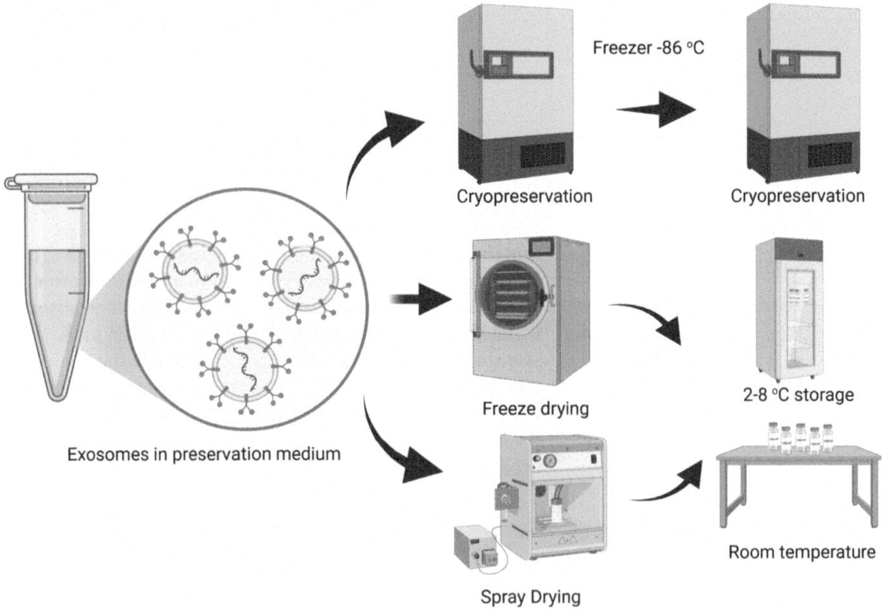

Fig. 4.1 Methods and conditions for exosome preservation. Exosomes can be suspended in a preservation medium containing a cryoprotectant reagent and stored at −86 °C. Alternatively, exosomes can undergo lipophilic processes such as freeze drying or spray drying and be stored at 2–8 °C or at room temperature for short-term preservation. (The figure is created using BioRender.com)

activity, and cryoprotectant agents can help withdraw water from the EVs. Similarly, in freeze drying, lyoprotectants are used in cryopreservation.

Exosomes can be preserved through freezing or lyophilization [29]; however, the preservation method must be carefully studied to ensure that the exosome products retain their structure and function. The use of preservatives (cryoprotectants or lyoprotectants) in the preservation process is necessary for both freezing and lyophilization. In clinical applications, the choice of cryopreservatives is determined by the effectiveness of the preservation process and by the toxicity of the preservatives toward humans. Products containing exosomes are often not washed to remove preservatives before use. At present, trehalose is considered the most effective preservative for both cryopreservation and lyophilization [30]. The optimal temperature for the storage of exosomes is −80°C.

Currently, due to the complexity and high cost of transporting frozen exosomes, companies producing exosome products employ processes based on freeze drying [31–33]. Freeze-dried products can be transported and stored more easily; studies have shown that these products can be stored at temperatures of 2–8 °C, or even at room temperature [34]. This advancement has ushered in a new era in the production and distribution of exosome products derived from stem cells for research and therapeutic purposes.

3 Current Preservation Strategies for EVs

3.1 Cryopreservation

3.1.1 Cryopreservation Medium

The cryopreservation medium for exosomes helps to maintain exosome morphology during freezing and thawing. Similar to cells, a cryopreservation medium can stabilize the osmolarity of the medium as well as prevent ice crystal formation inside the exosomes. During the freezing process, when the medium nearly reaches 0 °C, the concentration of soluble agents in the medium will increase the medium's osmolarity. Ice crystals form inside the exosomes when the temperature goes below 0 °C causing the exosome's membrane to break.

Therefore, cryoprotectants are added to the preservation media. Although several kinds of cryoprotectants can be used, DMSO, glycerol, and trehalose are the most common. DMSO (dimethyl sulfoxide) is an organosulfur compound that can freeze at 18.5 °C. It is widely used in the cryopreservation of mammalian cells as well as stem cells. This agent can penetrate the cells through the phospholipid membrane. Glycerol is another penetrating cryoprotectant agent that can form hydrogen bonds with water molecules. A mixture of glycerol and water can prevent the formation of ice crystals at temperatures as low as −37.8 °C. Trehalose is a disaccharide consisting of two molecules of glucose; in contrast to DMSO and glycerol, trehalose is a non-penetrating cryoprotectant. Trehalose has been successfully used to cryopreserve stem cells and mammalian cells.

Wu et al. (2015) showed that a cryopreservation medium containing 10% or 20% DMSO could maintain the morphology of exosomes; however, the DMSO-based medium could not preserve exosomal RNAs [35]. In patent number CN111088226A, exosomes from placenta mesenchymal stem cells are cryopreserved in 20% SR serum-free medium, 25–30% glycerol, and 15% vitamin C.

In 2016, Bosch et al. showed that the addition of 25 mM trehalose in PBS could prevent aggregation and cryodamage of exosomes [30]. Cryopreservation with 25 mM of trehalose can help to maintain the morphology and functionality of exosomes. These preserved exosomes could support hematopoietic stem cell culture and expansion [27]. Ruzycka-Ayoush et al. (2023) successfully stored exosomes derived from A549 in PBS supplemented with 25 mM trehalose at −80 °C. Under these conditions, the concentration, size distribution, zeta potential, and total protein levels of cryopreserved exosomes remained constant [36].

3.1.2 Cryopreservation Temperature

Because proteins as well as nucleic acids can be degraded at room temperature, exosomes should be frozen at ultralow temperatures. In 2011, Sokolova et al. applied nanoparticle tracking analysis (NTA) to measure the size of exosomes at different temperatures and suggested that at 37 °C exosome size is reduced more than at 4 °C [37]. Lee et al. (2016) compared exosomal proteins, RNA, and exosome markers from exosomes stored at −70 °C and room temperature after 10 days [38]. The results showed that the amounts of both proteins and RNA in the exosomes were reduced when stored at RT compared to −70 °C and 4 °C. The CD63 surface markers of exosomes were also significantly reduced after 10 days of storage at RT. The size of exosomes was altered as detected by flow cytometry with dispersed populations when the samples were stored at RT for 10 days [38].

Cheng et al. (2019) performed a comprehensive study concerning the effects of storage conditions of exosomes [39]. They stored exosomes at temperatures of −80 °C, −20 °C, 4 °C, 37 °C, and 60 °C (through 1–5 cycles of freezing to −80 °C and thawing). Their results confirmed that the concentration of exosomes decreased under storage at 37 °C and 60 °C as well as at −20 °C and −80 °C compared to 4 °C. This means that the higher temperature and freezing-thawing cycles can destroy exosomal membranes [39]. Budgude et al. (2020) showed that MVs and exosomes from stem cells can be kept at −80 °C in PBS supplemented with trehalose. Under these conditions, exosomes maintained their structural integrity and their biological properties [27]. To sum up, cryopreservation temperatures can directly affect the exosome structure. Long-term storage of exosomes should be done at temperatures below −70 °C.

3.1.3 Freezing Methods

As the cryopreservation of exosomes involves cell freezing, the protocols usually employ a freezing rate of 1–2 °C/min up to −86 °C. Then, exosomes can be stored at −86 °C or −196 °C. The slow freezing aims to prevent the formation of ice crystals that can damage the exosomes. Similar to cell thawing, cryopreserved exosomes can be rapidly thawed, when vials or bottles are placed in a 37 °C water bath.

Vitrification is another method used to cryopreserve exosomes. The method involves rapidly cooling the exosomes to subzero temperatures. The common protocol involves placing the exosomes in the cryoprotectants into liquid nitrogen. This method also can prevent the formation of ice crystals in the exosomes. To reduce the amount of water inside exosomes, a high concentration of a CPA (e.g., DMSO or glycerol) is used in vitrification. The exosomes are loaded into cryotubes and rapidly cooled to −196 °C before being immersed in liquid nitrogen.

3.2 *Freeze Drying*

3.2.1 Principles of Freeze Drying

Freeze drying is a method used in exosome preservation that involves removing the water from the exosomes while preserving their structure and function. The process begins by freezing the exosomes. The exosomes are then placed in a vacuum chamber where the pressure is reduced and the temperature is slowly increased. This causes the water in the exosomes to evaporate, leaving behind a dry powder that can be stored for long-term preservation. The exosomes are then sealed in a container to prevent any further moisture from entering. Freeze drying helps to preserve the structure and function of the exosomes, allowing them to be used in research and therapeutic applications.

The typical lyophilization process consists of three steps. The first step is freezing, where the liquid in the sample is frozen. This step is performed by placing the solution containing the EVs in a freezer compartment at −40 to −80 °C for several hours. The second step is primary drying or ice sublimation; in this step, the water in the solid state is converted into gas without passing through the liquid state. Generally, during the second step, the temperature is maintained at −40°°C under vacuum pressure. The third step involves raising the temperature to dry the sample. During this step, the temperature is normally increased to 20°°C.

3.2.2 Lyoprotectants

In almost all protocols, exosomes are resuspended in the medium with trehalose. In a recent study, Lyu et al. (2023) succeeded in storing extracellular vesicles from human corneal stromal stem cells by dry-freezing. In their study, trehalose was dissolved in PBS at a final concentration of 50 mM trehalose. EVs were frozen using liquid nitrogen. After 30 min, the samples were dried in a vacuum overnight and stored at RT for 4 weeks. The lyophilized EVs were rehydrated with distilled water. The rehydrated EVs stored at RT retained their in vitro biological activities [40].

Baradie et al. (2020) combined 50 mM trehalose and 2.5% PVP40 to dry-freeze exosomes from adipose-derived stem cells. The samples were frozen at −80 °C overnight. Then, samples are placed in the freezer dryer for 10 h. The products are stored at room temperature in a tightly sealed box to prevent moisture absorption and light exposure. The lyophilized EVs are rehydrated with ddH_2O [41]. These rehydrated EVs can be used to prevent hypoxia-induced muscle cell injury [41].

Trehalose is a nontoxic agent, and dry-freezing is a popular method used to store proteins, liposomes, and nanoparticles in the pharmaceutical industry. Freeze drying is a promising method for exosome storage in clinical applications.

3.3 Spray Drying

Spray drying is a suitable method for preserving various biological molecules such as vaccines, proteins, and peptides [42–45]. Recently, this method has been tested for the drying of exosomes as well as cell-secreted vesicles. In principle, the spray drying process proceeds as follows: a solution containing exosomes is sprayed into an environment with hot air, causing the solution droplets containing exosomes to be rapidly dehydrated and form a powder. A typical spray drying process consists of five steps [1]. Concentration: The solution containing the secretory vesicles is concentrated before being fed into the spray dryer [2]. Atomization: the solution is atomized to create optimal conditions for evaporation [3]. Droplet-air contact: in the chamber, the droplets containing secretory vesicles are atomized and come into contact with hot air. Up to 95% of the water contained in the droplets evaporates within seconds [4]. Droplet drying: water evaporation occurs in two stages. In the first stage, there is enough moisture in the droplet to replace the liquid evaporated at the surface, and evaporation occurs at a relatively constant rate. In the second stage, evaporation takes place when there is no longer enough moisture to maintain saturation at the surface of the droplet, leading to the formation of a dry shell around the droplet's surface [5]. Separation: this is the final stage of the spray drying process. Compared to the freeze-drying method, the spray-drying method is faster and simpler, resulting in lower total costs.

4 Conclusion

Preservation is an important aspect of exosome technology. Preservation allows exosomes to be stored and easily transported. Although exosomes are not cells, they carry crucial information in the form of nucleic acids and proteins. Damage to exosomes can alter their biological activity. Currently, there are three main types of processes for preserving exosomes: freezing, lyophilization (freeze drying), and spray drying. Among these, the freezing and lyophilization processes have been studied on various types of exosomes. Most such studies concerning freezing and lyophilization have affirmed that using cryoprotectants during freezing and lyophilization is necessary to preserve the structure and function of exosomes.

Competing Interests The authors declare that they have no competing interests.

References

1. Gurung S, Perocheau D, Touramanidou L, Baruteau J. The exosome journey: from biogenesis to uptake and intracellular signalling. Cell Commun Signal. 2021;19(1):47.
2. Isaac R, Reis FCG, Ying W, Olefsky JM. Exosomes as mediators of intercellular crosstalk in metabolism. Cell Metab. 2021;33(9):1744–62.
3. Deng S, Cao H, Cui X, Fan Y, Wang Q, Zhang X. Optimization of exosome-based cell-free strategies to enhance endogenous cell functions in tissue regeneration. Acta Biomater. 2023;171:68–84.
4. Muthu S, Bapat A, Jain R, Jeyaraman N, Jeyaraman M. Exosomal therapy-a new frontier in regenerative medicine. Stem Cell Investig. 2021;8:7.
5. Chen YF, Luh F, Ho YS, Yen Y. Exosomes: a review of biologic function, diagnostic and targeted therapy applications, and clinical trials. J Biomed Sci. 2024;31(1):67.
6. Huang B, Lu J, Ding C, Zou Q, Wang W, Li H. Exosomes derived from human adipose mesenchymal stem cells improve ovary function of premature ovarian insufficiency by targeting SMAD. Stem Cell Res Ther. 2018;9(1):216.
7. Li Z, Zhang M, Zheng J, Tian Y, Zhang H, Tan Y, et al. Human umbilical cord mesenchymal stem cell-derived exosomes improve ovarian function and proliferation of premature ovarian insufficiency by regulating the hippo signaling pathway. Front Endocrinol (Lausanne). 2021;12:711902.
8. Chi PL, Cheng CC, Wang MT, Liao JB, Kuo SH, Lin KC, et al. Induced pluripotent stem cell-derived exosomes attenuate vascular remodelling in pulmonary arterial hypertension by targeting HIF-1α and Runx2. Cardiovasc Res. 2024;120(2):203–14.
9. Tan W, Li Y, Ma L, Fu X, Long Q, Yan F, et al. Exosomes of endothelial progenitor cells repair injured vascular endothelial cells through the Bcl2/Bax/Caspase-3 pathway. Sci Rep. 2024;14(1):4465.
10. Yuan F, Peng W, Yang Y, Xu J, Liu Y, Xie Y, et al. Endothelial progenitor cell-derived exosomes promote anti-inflammatory macrophages via SOCS3/JAK2/STAT3 axis and improve the outcome of spinal cord injury. J Neuroinflammation. 2023;20(1):156.
11. Madhu LN, Kodali M, Upadhya R, Rao S, Somayaji Y, Attaluri S, et al. Extracellular vesicles from human-induced pluripotent stem cell-derived neural stem cells alleviate proinflammatory cascades within disease-associated microglia in Alzheimer's disease. J Extracell Vesicles. 2024;13(11):e12519.

12. Wang C, Cheng F, Han Z, Yan B, Liao P, Yin Z, et al. Human-induced pluripotent stem cell-derived neural stem cell exosomes improve blood-brain barrier function after intracerebral hemorrhage by activating astrocytes via PI3K/AKT/MCP-1 axis. Neural Regen Res. 2025;20(2):518–32.
13. Zhang B, Lai RC, Sim WK, Choo ABH, Lane EB, Lim SK. Topical application of mesenchymal stem cell exosomes alleviates the Imiquimod induced psoriasis-like inflammation. Int J Mol Sci. 2021;22(2)
14. Ding C, Zou Q, Wang F, Wu H, Chen R, Lv J, et al. Human amniotic mesenchymal stem cells improve ovarian function in natural aging through secreting hepatocyte growth factor and epidermal growth factor. Stem Cell Res Ther. 2018;9(1):55.
15. Han M, Yang H, Lu X, Li Y, Liu Z, Li F, et al. Three-dimensional-cultured MSC-derived exosome-hydrogel hybrid microneedle Array patch for spinal cord repair. Nano Lett. 2022;22(15):6391–401.
16. Xin W, Qiang S, Jianing D, Jiaming L, Fangqi L, Bin C, et al. Human bone marrow mesenchymal stem cell-derived exosomes attenuate blood-spinal cord barrier disruption via the TIMP2/MMP pathway after acute spinal cord injury. Mol Neurobiol. 2021;58(12):6490–504.
17. Zhang C. Exosomes derived from mesenchymal stem cells: therapeutic opportunities for spinal cord injury. Bull Exp Biol Med. 2024;176(6):716–21.
18. Sojoudi K, Azizi H, Skutella T. A review of the potential of induced pluripotent stem cell-derived exosome as a novel treatment for male infertility. Biotechnol Genet Eng Rev. 2024;40(2):1353–78.
19. Liu X, Li Q, Niu X, Hu B, Chen S, Song W, et al. Exosomes secreted from human-induced pluripotent stem cell-derived mesenchymal stem cells prevent osteonecrosis of the femoral head by promoting angiogenesis. Int J Biol Sci. 2017;13(2):232–44.
20. Park GH, Kwon HH, Seok J, Yang SH, Lee J, Park BC, et al. Efficacy of combined treatment with human adipose tissue stem cell-derived exosome-containing solution and microneedling for facial skin aging: a 12-week prospective, randomized, split-face study. J Cosmet Dermatol. 2023;22(12):3418–26.
21. Wang T, Gao H, Wang D, Zhang C, Hu K, Zhang H, et al. Stem cell-derived exosomes in the treatment of melasma and its percutaneous penetration. Lasers Surg Med. 2023;55(2):178–89.
22. Xie X, Song Q, Dai C, Cui S, Tang R, Li S, et al. Clinical safety and efficacy of allogenic human adipose mesenchymal stromal cells-derived exosomes in patients with mild to moderate Alzheimer's disease: a phase I/II clinical trial. Gen Psychiatr. 2023;36(5):e101143.
23. Dave C, Mei SHJ, McRae A, Hum C, Sullivan KJ, Champagne J, et al. Comparison of freshly cultured versus cryopreserved mesenchymal stem cells in animal models of inflammation: a pre-clinical systematic review. elife. 2022:11.
24. Zhao AG, Shah K, Cromer B, Sumer H. Comparative analysis of extracellular vesicles isolated from human mesenchymal stem cells by different isolation methods and visualisation of their uptake. Exp Cell Res. 2022;414(2):113097.
25. Ahmadian S, Jafari N, Tamadon A, Ghaffarzadeh A, Rahbarghazi R, Mahdipour M. Different storage and freezing protocols for extracellular vesicles: a systematic review. Stem Cell Res Ther. 2024;15(1):453.
26. Susa F, Limongi T, Borgione F, Peiretti S, Vallino M, Cauda V, et al. Comparative studies of different preservation methods and relative freeze-drying formulations for extracellular vesicle pharmaceutical applications. ACS Biomater Sci Eng. 2023;9(10):5871–85.
27. Budgude P, Kale V, Vaidya A. Cryopreservation of mesenchymal stromal cell-derived extracellular vesicles using trehalose maintains their ability to expand hematopoietic stem cells in vitro. Cryobiology. 2021;98:152–63.
28. Lo HL, Lin SY, Ho CJ, Ming-Kung Y, Lu CC. Effect of lyophilized exosomes derived from umbilical cord stem cells on chronic anterior cruciate ligament cell injury. J Orthop Surg Res. 2024;19(1):554.

29. Zhang Y, Bi J, Huang J, Tang Y, Du S, Li P. Exosome: a review of its classification, isolation techniques, storage, diagnostic and targeted therapy applications. Int J Nanomedicine. 2020;15:6917–34.
30. Bosch S, de Beaurepaire L, Allard M, Mosser M, Heichette C, Chrétien D, et al. Trehalose prevents aggregation of exosomes and cryodamage. Sci Rep. 2016;6(1):1–11.
31. Guarro M, Suñer F, Lecina M, Borrós S, Fornaguera C. Efficient extracellular vesicles freeze-dry method for direct formulations preparation and use. Colloids Surf B Biointerfaces. 2022;218:112745.
32. El Baradie KBY, Nouh M, O'Brien Iii F, Liu Y, Fulzele S, Eroglu A, et al. Freeze-dried extracellular vesicles from adipose-derived stem cells prevent hypoxia-induced muscle cell injury. Front Cell Dev Biol. 2020;8:181.
33. Paresishvili T, Kakabadze Z. Freeze-dried mesenchymal stem cells: from bench to bedside. Review Adv Biol (Weinh). 2024;8(2):e2300155.
34. Charoenviriyakul C, Takahashi Y, Nishikawa M, Takakura Y. Preservation of exosomes at room temperature using lyophilization. Int J Pharm. 2018;553(1–2):1–7.
35. Wu Y, Deng W, Klinke DJ II. Exosomes: improved methods to characterize their morphology, RNA content, and surface protein biomarkers. Analyst. 2015;140(19):6631–42.
36. Ruzycka-Ayoush M, Nowicka AM, Kowalczyk A, Gluchowska A, Targonska A, Mosieniak G, et al. Exosomes derived from lung cancer cells: isolation, characterization, and stability studies. Eur J Pharm Sci. 2023;181:106369.
37. Sokolova V, Ludwig AK, Hornung S, Rotan O, Horn PA, Epple M, et al. Characterisation of exosomes derived from human cells by nanoparticle tracking analysis and scanning electron microscopy. Colloids Surf B Biointerfaces. 2011;87(1):146–50.
38. Lee M, Ban J-J, Im W, Kim M. Influence of storage condition on exosome recovery. Biotechnol Bioprocess Eng. 2016;21:299–304.
39. Cheng Y, Zeng Q, Han Q, Xia W. Effect of pH, temperature and freezing-thawing on quantity changes and cellular uptake of exosomes. Protein Cell. 2019;10(4):295–9.
40. Lyu N, Knight R, Robertson SY, Dos Santos A, Zhang C, Ma C, et al. Stability and function of extracellular vesicles derived from immortalized human corneal stromal stem cells: a proof of concept study. AAPS J. 2022;25(1):8.
41. El Baradie KBY, Nouh M, O'Brien F III, Liu Y, Fulzele S, Eroglu A, et al. Freeze-dried extracellular vesicles from adipose-derived stem cells prevent hypoxia-induced muscle cell injury. Front Cell Dev Biol. 2020;8:181.
42. Bahr MM, Amer MS, Abo-El-Sooud K, Abdallah AN, El-Tookhy OS. Preservation techniques of stem cells extracellular vesicles: a gate for manufacturing of clinical grade therapeutic extracellular vesicles and long-term clinical trials. Int J Vet Sci Med. 2020;8(1):1–8.
43. Janouskova O, Herma R, Semeradtova A, Poustka D, Liegertova M, Hana Auer M, et al. Conventional and nonconventional sources of exosomes-isolation methods and influence on their downstream biomedical application. Front Mol Biosci. 2022;9:846650.
44. Kusuma GD, Barabadi M, Tan JL, Morton DAV, Frith JE, Lim R. To protect and to preserve: novel preservation strategies for extracellular vesicles. Front Pharmacol. 2018;9:1199.
45. Pinto JT, Faulhammer E, Dieplinger J, Dekner M, Makert C, Nieder M, et al. Progress in spray-drying of protein pharmaceuticals: literature analysis of trends in formulation and process attributes. Dry Technol. 2021;39(11):1415–46.

Chapter 5
Manufacturing and Quality Control of Mesenchymal Stem Cell–Derived Exosomes for Clinical Application

Ngoc Bich Vu, Phuong Thi-Bich Le, and Phuc Van Pham

1 Introduction

Mesenchymal stem cells (MSCs) have been studied for several years. This type of cell was discovered in the 1950s, with the first report of its presence through the process of bone marrow transplantation in 1956 [1]. Friedenstein et al. were the first to determine the osteogenic differentiation potential of bone marrow stem cell transplantations [2]. By 1980, these cells were also shown to have the ability to differentiate into cartilage and bone [3]. The concept of MSCs was introduced by Dr. Arnold Caplan in 1991 in a study that described a special technique for isolating stem cells from the bone marrow [4].

Currently, MSCs are one of the most important stem cell types in research and applications [5]. They are the most clinically used cells to date, owing to their superior properties in disease treatment. For example, they possess immunomodulatory and anti-inflammatory capabilities, stimulate angiogenesis, inhibit the aging process and apoptosis of other cells, and resist fibrotic scarring [6–10]. Importantly, they express low levels of HLA surface antigens and have low immunogenicity, allowing them to be used in allogeneic transplantation without the need for HLA matching between mesenchymal stem cells and the recipient [11–13]. As a result, they have been used to produce therapeutic drugs for many diseases and have been approved by health authorities in several countries [14–17].

N. B. Vu · P. V. Pham (✉)
VNUHCM-US Stem Cell Institute, University of Science, Viet Nam National University Ho Chi Minh City, Ho Chi Minh, Vietnam
e-mail: ngocvu@sci.edu.vn; phucpham@sci.edu.vn

P. T.-B. Le
Stem Cell Unit, Van Hanh General Hospital, Ho Chi Minh City, Vietnam

P. V. Pham, S. Fagoonee (eds.), *Stem Cell-Derived Exosomes and Microvesicles*, Stem Cells in Clinical Applications, https://doi.org/10.1007/978-3-031-90698-5_5

Despite their many advantages and potential, MSCs still have some limitations that prevent their widespread use and value for patients. For instance, transporting cells is challenging because of the need to store them at very low temperatures, and the cell quality can easily change during transportation if the temperature is not maintained. Moreover, the regulatory process for approving these products is complex because of their cellular nature. Hence, secretions or secretomes of MSCs, including exosomes, are emerging as new candidates [18, 19].

Exosomes are secreted from cells in general and MSCs in particular. These vesicles are 30–150 nm in size and carry crucial information for cell-to-cell communication. This information is encoded by nucleic acid or protein molecules within the exosome. These molecules can elicit biological effects similar to those of MSCs, such as immunomodulation, anti-inflammation, scar inhibition, and angiogenesis [20, 21]. Therefore, exosomes are gradually becoming a potential alternative to MSCs. Using exosomes for disease treatment or supportive therapy is thought to have advantages over MSCs, such as non-cellular nature, reduced cellular graft-related risks, easier product transport, especially at cool or even room temperatures, and more convenient storage conditions. These factors have accelerated the research and development of exosome products, leading to significant achievements in recent years.

This chapter summarizes and introduces the main aspects of the process of manufacturing and controlling the quality of exosomes from human MSCs for clinical use. This chapter details the production process, from preparing MSCs for the purification and collection of exosomes (Fig. 5.1). Control points and the necessary quality criteria to ensure safety and effectiveness in clinical use are presented in this chapter.

2 Mesenchymal Stem Cells–Derived Exosome Manufacturing

2.1 Preparation of Mesenchymal Stem Cells

The source of stem cells used in exosome production must be carefully evaluated to meet important quality requirements. The following issues need to be considered when using MSCs for exosome production.

The tissue source for stem cell isolation must be approved by an ethics committee to ensure compliance with other relevant legal requirements.

Tissue donors for MSC collection must be screened to meet the minimum safety standards for tissue samples. Donors must be screened to confirm that they are not infected with Human Immunodeficiency Virus (HIV), Hepatitis B Virus (HBV), Hepatitis C Virus (HCV), Human T-lymphotropic Virus (HTLV), Epstein-Barr Virus (EBV), Human Cytomegalovirus (HCMV), Treponema pallidum (TP). In addition, information related to medical and family history of genetic disorders, cardiovascular diseases, and cancer must also be screened.

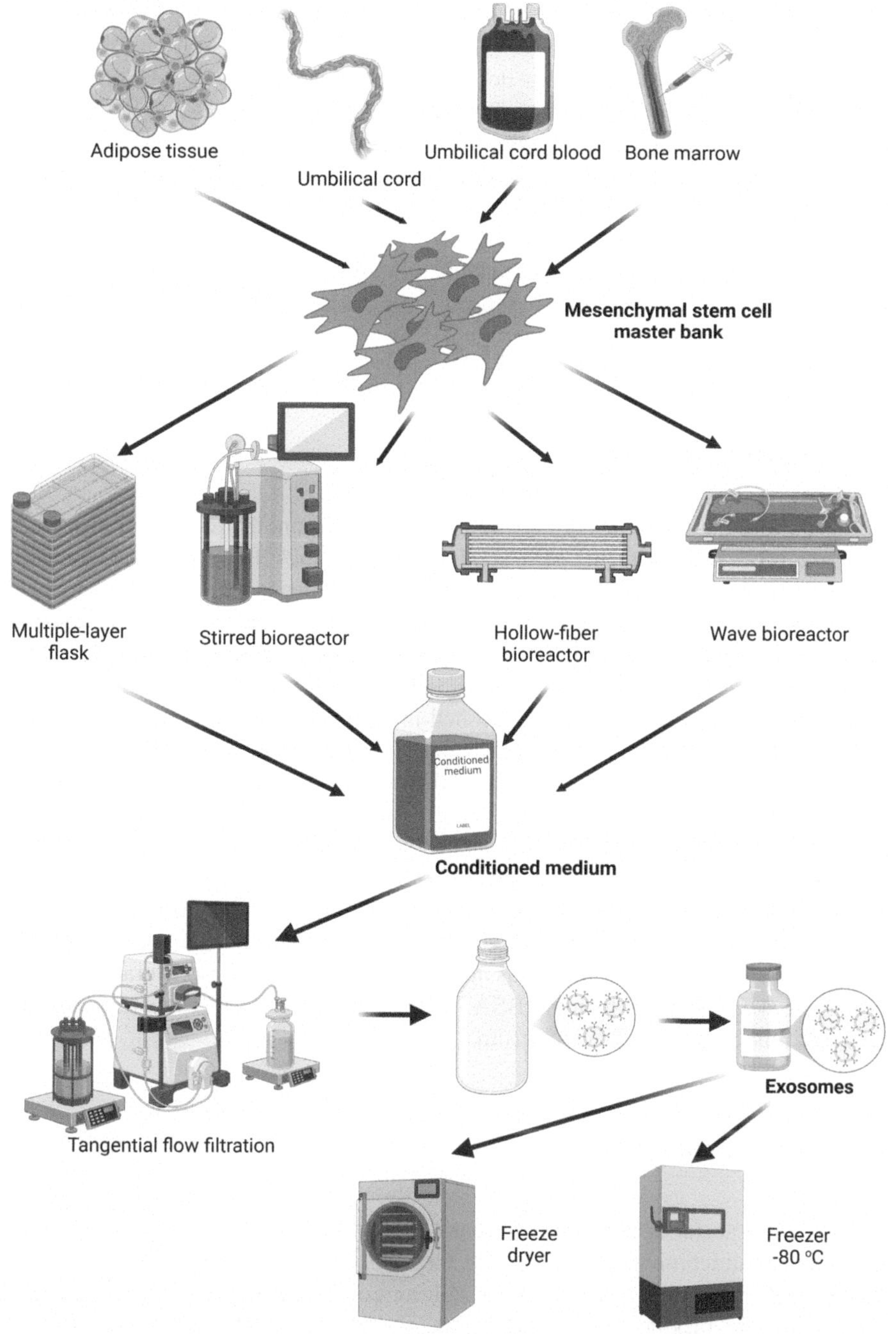

Fig. 5.1 The flowchart of mesenchymal stem cell–derived exosome manufacturing. First, mesenchymal stem cells should be isolated from adipose tissue, bone marrow, umbilical cord, or umbilical cord blood to prepare the master cell bank. In a batch of exosomes, the mesenchymal stem cells from the master cell bank are used to expand in bioreactors or multilayer flasks to obtain conditioned medium. Exosomes from the conditioned medium are collected by tangential flow filtration with a molecular weight cut-off of 100–300 kDa. The obtained exosomes can be preserved by cryopreservation or freeze-drying

The MSC source isolated from donated tissue must meet the necessary standards, including satisfying the phenotypic requirements of MSCs and being tested and confirmed to be free from bacteria, fungi, and mycoplasma. Moreover, to ensure the proliferative capacity of MSCs, it is important to limit the senescence-associated secretory phenotypes (SASPs) that appear in the culture medium and ensure the quality of secreted exosome vesicles. MSCs used in the production process are typically carefully checked for senescent status, or cells from passages 4 or 5 are used. In some cases, the MSC source may need to be karyotyped before being used for exosome production.

2.2 Culture Medium

Various culture media are used, depending on the different technological processes during the process of culturing MSCs to harvest exosomes. In most cases, to ensure the effectiveness of the harvest, media with essential nutrients and proteins, including serum, are used during the initial period when culturing cells at low density. When the cell density reached 50% confluency, the cells were washed with saline, and an exosome harvest medium was added. Typically, cells are cultured in this medium for 24–48 h to collect conditioned medium. Conditioned medium should be collected every 48 h from MSCs showing 50–90% confluency [22].

Next, the culture medium was collected to isolate the exosomes. The exosome harvest medium must meet certain criteria: it should not contain exogenous secretory vesicles or supplementary components, such as bovine serum, platelet lysate, pituitary extracts, bile salts, and several other elements. The exosome harvest medium did not contain phenol red or antimicrobial antibiotics. To ensure that the exosome products were not contaminated with bovine serum albumin, a medium supplemented with bovine serum was not used during the MSC proliferation phase. It is essential to check for residual BSA in the final preparation, if it is necessary to use a medium containing bovine serum.

2.3 Culture Condition

MSCs require surfaces that are adherent to substrates. Therefore, culturing these cells requires attachment sites for MSCs. Generally, MSCs survive at 37 °C and 5% CO_2. However, in some cases, hypoxic culture conditions (5–10% O_2) are used to create exosome products with certain special biological characteristics. Recent studies have shown that the ability of MSCs to secrete exosomes can be enhanced under various culture conditions. However, under special conditions, the characteristics of MSCs may change; therefore, the biological properties of the exosomes they secrete may not be the same as under normal conditions. This necessitates careful evaluation or assessment of the biological properties of exosomes produced under different conditions before use.

2.4 Scale-Up

The scale-up of MSC expansion to collect secretions is commonly performed in bioreactors. Because MSCs need to adhere to a substrate to survive and proliferate, in typical bioreactor systems, these cells attach to a surface substrate. In some systems, MSCs are cultured to adhere to the surface of beads suspended in the culture medium. To prevent the beads from settling [23], stirring or agitation was used to mix the medium. In other systems, substrate surfaces are made of special materials to assist in cell adhesion (such as hollow fibers). Some studies have suggested that exosomes harvested from cultures in bioreactors exhibit stronger biological activity than those cultured in flasks [24–26].

However, optimizing the culture conditions for MSCs in bioreactors presents several challenges [27]. Therefore, some entities continue to use MSC cultures in large or multilayer flasks. This system is often referred to as a scaled-out system. Currently, some multi-layer flasks can provide surface areas for MSCs to adhere to, reaching up to thousands of square centimeters, making them a suitable solution for large-scale culture of MSCs for exosome harvesting.

2.5 Cell Culture Supernatant Collection

The collection of conditioned medium is an important step in exosome production. The most crucial aspect of this step is the selection of the appropriate time for harvesting. Typically, the appropriate time to harvest exosomes is 48 hours after the exosome-harvesting medium is added to the culture when the cell confluence reaches 50%. MSCs are considered the healthiest, and the exosomes obtained are optimal in terms of both quality and quantity.

However, in some currently available processes, exosomes are collected throughout the MSC culture. This means that the cell culture medium can still be used for exosome isolation during proliferation. This system often integrates a bioreactor with an exosome isolation system.

2.6 Exosome Isolation and Preservation

Exosome isolation involves various methods to obtain exosomes from cell culture supernatants. Although multiple techniques are available for harvesting exosomes from culture fluids, tangential flow filtration (TFF) is commonly used in practical production. This method is easy to implement, can operate on a large volume of cell secretions, is cost-effective, and ensures that purification is suitable for clinical applications [24]. To meet the demand for exosome production for research and clinical applications, many automatic TFF systems have been developed and utilized.

When using TFF systems to harvest exosomes, conditioned media are often prefiltered through membranes with a pore size of 0.22 μm to remove cell debris and secreted particles larger than 220 nm in diameter. However, membranes are prone to clogging, resulting in a significant loss of exosomes; thus, this filtration can be carried out after TFF for exosome harvesting. The pore size of the tangential flow membranes used to harvest exosomes is typically 100–300 kDa. The membranes retained vesicles larger than 30 nm in diameter. It is important to note that reusing filtration columns and lines can increase endotoxin levels during preparation if they are contaminated with microorganisms.

The simplest method for preserving exosome products from MSCs while maintaining their activity, which is freezing at −86 °C. Utilizing this preservation method is straightforward, involving suspending the harvested exosome quantity in a freezing medium containing a cryoprotectant, typically trehalose [28, 29]. The product can then be placed directly in a −86 °C freezer or slowly cooled at 1–2 °C per minute to −86 °C. MSC-derived exosomes can be lyophilized for preservation [30, 31]. This process has also been researched and published, with promising results. During lyophilization, exosomes are suspended in medium containing a preservative, usually trehalose. The product was frozen at −80 °C and then sublimated at −40 °C to gradually remove moisture. Finally, the product was dried by increasing the temperature to approximately 20 °C until it was completely dry. Lyophilized products can be stored for a short period at room temperature or longer at 2–8 °C.

2.7 Package, Storage, and Transport

The packaging of exosome products requires that it meets certain special requirements, such as the packaging material having no adsorption properties towards exosomes and not causing exosome aggregation. These materials are in direct contact with the product, so they must be sterile and pyrogen-free. Product usage instructions must be placed inside a pouch containing the product. For products that need to be stored at −86 °C, the material containing the product must withstand this temperature.

The exosome product label must provide all the necessary information, including the product's specific name, the number of exosomes contained within the product, the cellular origin used to produce the exosomes, the volume of the product, the product's manufacturing date, the production batch number, storage conditions, and expiration date. Each product must have a quality certificate, according to its production batch.

Typically, products containing exosomes from MSCs are stored at −80 °C [32]. The product should not be repeatedly thawed and refrozen, as this can affect its internal quality. With new technological processes, these products can be stored in a lyophilized form. However, the shelf life of the product must be carefully evaluated and controlled to ensure safety and effectiveness.

Products may be transported under dry ice conditions if they need to be stored at −80 °C. Products stored in the lyophilized form should be transported at a temperature of 2–8 °C.

3 Quality Control of Exosomes for Clinical Applications

3.1 Morphology

Most exosomes have a cup-round or round membranous vesicle form. To evaluate their shape, they were observed under a transmission electron microscope (TEM).

3.2 Size

The size of exosome vesicles typically ranges from 30 to 150 nm. Size assessment can be performed using nanoparticle tracking analysis (NTA) or nano-flow analysis methods. When using these techniques, the peak distribution of the particles should be within the range 30–150 nm. Although the TEM technique can measure the size of exosome particles, it does not comprehensively show the distribution of particle sizes within the product; therefore, it is not used for this criterion.

3.3 Markers

To confirm that the vesicles in the product are exosomes, they must meet the following protein expression profile: positive for two out of three markers, CD9, CD63, and CD81; positive for one of the two markers, Alix and Tsg101; and negative for three markers, calnexin, histone 3, and GM130. The expression of these proteins was evaluated using flow cytometry. The positivity rate of positive markers should be above 90%, whereas the negativity rate of negative markers should be below 5%. These rates may vary depending on the regulations and standards of the different organizations and countries.

3.4 Granule Protein Ratio

The ratio of exosome particles to protein concentration was not lower than 10^8 particles/μg.

3.5 Microorganisms

The product must be negative for fungi, bacteria, mycoplasma, HIV, HBV, HCV, HTLV, EBV, HCMV, TP, and exogenous viral factors.

3.6 Endotoxin

Endotoxins must be less than 0.5 EU/mL.

3.7 Potency

The potency of a product is an important factor in determining its efficacy when used clinically. However, depending on the intended use, a product can be evaluated using several different assessments or potencies. For example, a product used to stimulate new blood vessel formation must be evaluated for its ability to promote neovascularization in vitro or in vivo.

4 Monitoring and Auditing

Monitoring and supervising the production process of products are necessary to ensure their quality. Generally, the production of exosome products is carried out in batches. In each batch, MSCs were from the same source, and the cells were cultured under the same conditions. In addition, purification, packaging, and storage processes should be integrated.

During production, from the preparation of MSCs to storage and distribution, multiple quality control checkpoints should be utilized rather than focusing solely on the final products. For instance, quality control of MSCs should be conducted before their use in cell culture production, the cell culture supernatant should be controlled before exosome isolation, exosomes should undergo quality control before preservation, and the finished product should be quality controlled. In a production batch with multiple packaging units, finished products must be randomly selected at a rate specified by each country for evaluation. For example, if product standards must be assessed for 3% of products, then three products out of a batch of 100 should be randomly selected for evaluation. All three products must meet the output criteria for the batch to be certified as standardized. A certificate of analysis and production process documentation (lot document) should be accompanied by each batch.

Storing products from production batches is essential for retrospective evaluation. Typically, the quantity of stored products should be sufficient for at least two retrospective assessments. The storage duration should be at least 1 year after the expiration date of the product.

5 Conclusion

Exosomes from MSCs are becoming therapeutic products of wide interest, alongside MSCs. With many advantages, exosomes from MSCs have been studied, evaluated, and used in treatment and beauty applications in many countries. Although no exosome products have been officially approved by the health authorities of any country as therapeutic drugs, clinical studies have shown that these are effective products in the treatment and support of various diseases. Therefore, the development of research and production processes for exosomes from MSCs has been implemented and has shown many positive results. This chapter provides comprehensive and specific information on the production process and quality control to ensure that exosome products can be used clinically.

Competing Interests The authors declare that they have no competing interests.

References

1. Danis A. Study of ossification in bone marrow grafts [Doctoral thesis]. Brussels: Université libre de Bruxelles, Faculty of Medicine; 1956.
2. Friedenstein AJ, Piatetzky S II, Petrakova KV. Osteogenesis in transplants of bone marrow cells. J Embryol Exp Morphol. 1966;16(3):381–90.
3. Ashton BA, Allen TD, Howlett CR, Eaglesom CC, Hattori A, Owen M. Formation of bone and cartilage by marrow stromal cells in diffusion chambers in vivo. Clin Orthop Relat Res. 1980;151:294–307.
4. Caplan AI. Mesenchymal stem cells. J Orthop Res. 1991;9(5):641–50.
5. Vizoso FJ, Costa LA, Eiro N. Mesenchymal stem cells and their derived products in ageing and diseases. Int J Mol Sci. 2024;25(13):6979.
6. Ryu JS, Jeong EJ, Kim JY, Park SJ, Ju WS, Kim CH, et al. Application of mesenchymal stem cells in inflammatory and fibrotic diseases. Int J Mol Sci. 2020;21(21):8366.
7. Martínez-Zarco BA, Jiménez-García MG, Tirado R, Ambrosio J, Hernández-Mendoza L. Mesenchymal stem cells: therapeutic option in ARDS, COPD, and COVID-19 patients. Rev Alerg Mex. 2023;70(2):89–101.

8. Shang L, Shao J, Ge S. Immunomodulatory functions of oral mesenchymal stem cells: novel force for tissue regeneration and disease therapy. J Leukoc Biol. 2021;110(3):539–52.
9. Liu Q, Zhang X, Zhu T, Xu Z, Dong Y, Chen B. Mitochondrial transfer from mesenchymal stem cells: mechanisms and functions. Mitochondrion. 2024;79:101950.
10. Wang J, Chen Z, Sun M, Xu H, Gao Y, Liu J, et al. Characterization and therapeutic applications of mesenchymal stem cells for regenerative medicine. Tissue Cell. 2020;64:101330.
11. Kot M, Baj-Krzyworzeka M, Szatanek R, Musiał-Wysocka A, Suda-Szczurek M, Majka M. The importance of HLA assessment in "off-the-shelf" allogeneic mesenchymal stem cells based-therapies. Int J Mol Sci. 2019;20(22):5680.
12. Li Y, Jin M, Guo D, Shen S, Lu K, Pan R, et al. Unveiling the immunogenicity of allogeneic mesenchymal stromal cells: challenges and strategies for enhanced therapeutic efficacy. Biomed Pharmacother. 2024;180:117537.
13. Schu S, Nosov M, O'Flynn L, Shaw G, Treacy O, Barry F, et al. Immunogenicity of allogeneic mesenchymal stem cells. J Cell Mol Med. 2012;16(9):2094–103.
14. Najima Y, Ohashi K. Mesenchymal stem cells: the first approved stem cell drug in Japan. 日本造血細胞移植学会雑誌. 2017;6(3):125–32.
15. Kurtzberg J, Abdel-Azim H, Carpenter P, Chaudhury S, Horn B, Mahadeo K, et al. A phase 3, single-arm, prospective study of Remestemcel-L, ex vivo culture-expanded adult human mesenchymal stromal cells for the treatment of pediatric patients who failed to respond to steroid treatment for acute graft-versus-host disease. Biol Blood Marrow Transplant. 2020;26(5):845–54.
16. Song J-S, Hong K-T, Kim N-M, Park H-S, Choi N-H. Human umbilical cord blood-derived mesenchymal stem cell implantation for osteoarthritis of the knee. Arch Orthop Trauma Surg. 2020;140:503–9.
17. Van Pham P. Stem cell drugs: the next generation of pharmaceutical products. Biomed Res Ther. 2016;3:1–15.
18. Rao D, Huang D, Sang C, Zhong T, Zhang Z, Tang Z. Advances in mesenchymal stem cell-derived exosomes as drug delivery vehicles. Front Bioeng Biotechnol. 2021;9:797359.
19. Shi H, Yang Z, Cui J, Tao H, Ma R, Zhao Y. Mesenchymal stem cell-derived exosomes: a promising alternative in the therapy of preeclampsia. Stem Cell Res Ther. 2024;15(1):30.
20. Tang Y, Zhou Y, Li HJ. Advances in mesenchymal stem cell exosomes: a review. Stem Cell Res Ther. 2021;12(1):71.
21. Maqsood M, Kang M, Wu X, Chen J, Teng L, Qiu L. Adult mesenchymal stem cells and their exosomes: sources, characteristics, and application in regenerative medicine. Life Sci. 2020;256:118002.
22. Börger V, Staubach S, Dittrich R, Stambouli O, Giebel B. Scaled isolation of mesenchymal stem/stromal cell-derived extracellular vesicles. Curr Protoc Stem Cell Biol. 2020;55(1):e128.
23. Wang X, Ouyang L, Chen W, Cao Y, Zhang L. Efficient expansion and delayed senescence of hUC-MSCs by microcarrier-bioreactor system. Stem Cell Res Ther. 2023;14(1):284.
24. Haraszti RA, Miller R, Stoppato M, Sere YY, Coles A, Didiot MC, et al. Exosomes produced from 3D cultures of MSCs by tangential flow filtration show higher yield and improved activity. Mol Ther. 2018;26(12):2838–47.
25. Yan L, Wu X. Exosomes produced from 3D cultures of umbilical cord mesenchymal stem cells in a hollow-fiber bioreactor show improved osteochondral regeneration activity. Cell Biol Toxicol. 2020;36(2):165–78.
26. Lin G, Pan W, He Y, Yi X, Zhou P, Lu J. Opportunities and perspectives for three dimensional culture of mesenchymal stem cell-derived exosomes. Tissue Eng Part B Rev. 2024;30(4):462–76.
27. Fernández-Santos ME, Garcia-Arranz M, Andreu EJ, García-Hernández AM, López-Parra M, Villarón E, et al. Optimization of mesenchymal stromal cell (MSC) manufacturing processes for a better therapeutic outcome. Front Immunol. 2022;13:918565.
28. Bosch S, de Beaurepaire L, Allard M, Mosser M, Heichette C, Chrétien D, et al. Trehalose prevents aggregation of exosomes and cryodamage. Sci Rep. 2016;6:36162.

29. Budgude P, Kale V, Vaidya A. Cryopreservation of mesenchymal stromal cell-derived extracellular vesicles using trehalose maintains their ability to expand hematopoietic stem cells in vitro. Cryobiology. 2021;98:152–63.
30. El Baradie KBY, Nouh M, O'Brien Iii F, Liu Y, Fulzele S, Eroglu A, et al. Freeze-dried extracellular vesicles from adipose-derived stem cells prevent hypoxia-induced muscle cell injury. Front Cell Dev Biol. 2020;8:181.
31. Qazi REM, Sajid Z, Zhao C, Hussain I, Iftikhar F, Jameel M, et al. Lyophilization based isolation of exosomes. Int J Mol Sci. 2023;24(13):10477.
32. Wu JY, Li YJ, Hu XB, Huang S, Xiang DX. Preservation of small extracellular vesicles for functional analysis and therapeutic applications: a comparative evaluation of storage conditions. Drug Deliv. 2021;28(1):162–70.

Chapter 6
New Advances in Stem Cell–Derived Extracellular Vesicle Functionalization Technology

Simone Tomasini, Maria Laura Centomo, Iva Neveux, and Riccardo Panella

1 Introduction

As discussed in this book, the growing evidence of the advantages of the therapeutic role of extracellular vesicles (EVs) derived from mesenchymal stromal/stem cells (MSC), hereinafter MSC-EVs, contributes to the strong rationale for developing EVs as an alternative cell-free therapeutic option. However, it should also be kept in mind that despite their success in preclinical studies, the use of MSC-EVs in clinical settings still faces several challenges. Specifically, several critical matters such as the production techniques, quantification and characterization, pharmacokinetics, targeting and transfer to the target sites, pharmacodynamics, and safety profile assessments need to be resolved to establish MSC-EVs as FDA-approved therapy. Lack of standardization in EV isolation methods and the large-scale production

S. Tomasini
Center for RNA Medicine, Department of Clinical Medicine, Aalborg University, Copenhagen, SV, Denmark
e-mail: sito@dcm.aau.dk

M. L. Centomo
Center for Genomic Medicine, Desert Research Institute, Reno, NV, USA

MD Anderson Cancer Center, Department of Molecular and Cellular Oncology, Division of Discovery Science, Houston, TX, USA
e-mail: mlcentomo@mdanderson.org

I. Neveux
Center for Genomic Medicine, Desert Research Institute, Reno, NV, USA

Department of Internal Medicine, University of Nevada, Reno, NV, USA
e-mail: ineveux@med.unr.edu

P. V. Pham, S. Fagoonee (eds.), *Stem Cell-Derived Exosomes and Microvesicles*, Stem Cells in Clinical Applications, https://doi.org/10.1007/978-3-031-90698-5_6

under current good manufacturing practice (cGMP) conditions to meet market demands are the major challenges. According to a worldwide survey by the International Society for Extracellular Vesicles (ISEV), several isolation methods with varying degrees of efficacy in terms of purity and quantity are now available [1, 2]. These include ultracentrifugation (UCF), differential centrifugation (DCF), size-exclusion chromatography (SEC), immune-based capturing (IBC), and precipitation methods with reagents like polyethylene glycol (PEG) [3, 4]. Interestingly, recent studies have demonstrated that EV isolation by UCF followed by SEC results in high yield while preserving functional and biophysical properties [5, 6]. However, applying these techniques to clinical platforms is not feasible because of difficulties in upscaling and automating such a process that do not facilitate a large-scale production. A promising approach to scale up the amount of isolated EV from MSCs should be the use of 3D culturing bioreactors [7] with no excessive cost. To ensure reliability, reproducibility, and massive EV production, several technical factors need to be standardized. Indeed, many cultural conditions can affect the quantity and quality of isolated MSC-EVs, such as cellular confluence, early versus later passage of cells, oxygen concentration, cytokines, heparin, and serum concentration [8]. Besides mass production and scalable methods for MSC-EV isolation, EV preservation is another important factor to consider in the clinical setting in order to maintain EV stability and integrity. To this end, trehalose has been proven to be a good cryoprotectant for cryogenic storage of clinical-grade MSC-EVs [9]. Though storage at −80 °C is a good option, it may come with cost limits and transportation difficulties. Therefore, efforts in freeze-drying techniques for storage at room temperature are moving ahead [10]. Fortunately, EVs compared to MSCs showed simplified cold chain reaction [11].

To be validated by FDA as drugs, pharmacokinetic and pharmacodynamic studies need to be performed. The EV biodistribution can be affected by dosage, route of administration, and cell source [12]. To reach the maximum therapeutic potential and establish the suitable therapeutic dosage, ADME studies are essential to better define adsorption, distribution, metabolism and excretion of specific formulations. It is well known that MSC-EVs express specific MSC receptors that facilitate the homing of injured tissues. However, the decoration of the EV surface with selective molecules provides bioengineered EVs with outstanding targeting ability. Despite

R. Panella (✉)
Center for RNA Medicine, Department of Clinical Medicine, Aalborg University, Copenhagen, SV, Denmark

Center for Genomic Medicine, Desert Research Institute, Reno, NV, USA

Resalis Therapeutics, Torino, Italy

Department of Neuroscience, Psychology, Drug Research and Child Health—NEUROFARBA—Pharmacology and Toxicology Section, University of Florence, Florence, Italy

European Biomedical Research Institute of Salerno (EBRIS), Salerno, Italy
e-mail: rpanella@dcm.aau.dk

all the limitations discussed above, MSC-EVs have moved to clinical studies rapidly. Preliminary results in terms of safety are encouraging based on their biocompatibility, low toxicity, and immunogenicity, shedding light on future studies.

Many of the clinical trials currently in progress use engineered EV rather than naïve EV to enhance the therapeutic potential. An engineering strategy is to incorporate bioactive exogenous molecules, including microRNAs (miRNAs) or small-molecule drugs, into the EVs through passive or active loading processes (Table 6.1) [19, 20]. Indeed, in a recent study, systemic injection of EV loaded with miRNA-124 was shown to promote neurogenesis in mouse models subjected to ischemic injury [21]. Accordingly, a phase 1–2 clinical trial (https://clinicaltrials.gov/ct2/show/NCT03384433) aims to investigate the safety and efficacy of genetically modified Allogenic MSC-EV enriched with miRNA-124 in ischemic stroke patients is currently in progress. Another strategy to improve the therapeutic effect is to engineer the EVs with targeting ability through specific surface modifications [20]. Although the dose and the route of administration influenced the EV biodistribution, once administered, EVs mainly accumulated in the liver, spleen, and intestine [12, 22]. Engineered EVs with targeting molecules increase their chance to reach the target tissue, thus avoiding clearance and reducing off-target effects. Therefore, lower therapeutic dosage will be possible while maintaining the therapeutic efficacy. Engineered EVs, discussed in the present chapter, hold a great potential to be

Table 6.1 Summary table including the most relevant methods for loading EVs with therapeutics

Methodology	Advantages	Disadvantages	References
Passive incubation with therapeutics	Noninvasive; no EV manipulation needed	The therapeutic must be hydrophobic; can require a chemical modification to the therapeutic compound	[13]
Electroporation	High efficiency; easy to scale up; GMP compatible; applicable to all chemistries	It can cause self-aggregation; can affect structural integrity	[14, 15]
Sonication	Applicable to hydrophilic molecules like nucleic acids; increases EV's loading capacity	Membrane disruption is problematic for production scale-up; it might affect surface modifications	[16]
Freeze-thaw cycles, extrusion, dialysis	Very flexible; creates big pores allowing the EV to be loaded with molecules with high molecular weight	Mechanical manipulation can disrupt membrane integrity and affect surface properties	[17]
Endogenous loading	Minimizes EV manipulation; allows harvesting from media EV already loaded with the desired molecule	Requires multiple troubleshooting steps involving cell manipulation; the therapeutic compound must be a molecule that has no cytotoxicity and can be synthetized by cells; difficult to control the amount of molecule loaded	[18]

rapidly translated into new effective strategies in the clinical setting and could become the treatment choice of the future.

2 Surface Modification

Due to their minuscule size, their ability to protect encapsulated drugs, pass through the brain blood barrier, [23–25] and enter avascular parts of a human body such as cartilage [26], EVs represent a promising drug delivery platform. EVs can deliver their content to target cells and the latter receive the functional genomic and proteomic content in various ways: the transmembrane receptors in the EVs can initiate receptor-ligand interactions and trigger a signaling cascade in the target cell [27]. EVs can also release their content into the recipient cell by internalization of the vesicle content either directly through the cell membrane or via endocytosis, [28–30] phagocytosis, or macropinocytosis. Other cellular entry routes of EVs include clathrin interactions, lipid raft interactions, and direct fusion [31]. Previous studies have indicated that the uptake of exosomes depends on time, temperature, and the type of recipient cells. For example, the uptake of exosomes by HEK 293 cells reaches saturation at about 12 h and decreases after 24 h [32]. Another study has shown that MSC-EVs and EVs from 3-day osteogenically differentiated MSCs had similar and robust internalization by human bone marrow–derived MSCs (hBMSCs) at 12 h [69]. Despite the above examples, the exact mechanisms by which EVs enter their target cells are not fully understood and examples available from literature are limited to specific types of EVs [33].

There are multiple routes of EV administration, and they vary according to disease characteristics. However, the optimal route for each specific disease has not yet been fully established. Organ-specific administrations, such as intracoronary route for cardiac diseases and intratracheal or inhalation route for lung disease, are viable alternatives to systemic infusion. Accordingly, topical application of allogenic MSC-EVs for rare genetic skin disorders is under progress in a phase 1-2a trial (https://clinicaltrials.gov/ct2/show/NCT04173650). It is worth noting that there are emerging engineering strategies for local or percutaneous administration through the combination of EVs with injectable biomaterials [34].

The major pitfall of using biologicals as drugs is poor delivery of the molecules to their target sites. Lipid nanoparticles (LNPs) offer some solution, but the recently approved FDA use of LNPs for treatment of transthyretin amyloidosis has its limitations. Despite the fact the LNPs protect their cargo and facilitate entry into cells, they display dose-limiting toxicity and deliver their cargo primarily into liver cells, delaying the delivery to other tissues [35]. Although the distribution of EVs is influenced by the dose and the route of administration, once administered, EVs mainly accumulate in the liver, spleen, and intestine [12, 22]. Therefore, researchers are striving to outfit EVs with new functionalities, such as tissue-specific targeting [20, 36]. EVs preferentially interact with their target cells, but these interactions are not very effective in vivo. Surface engineering offers solution to this hurdle, introducing

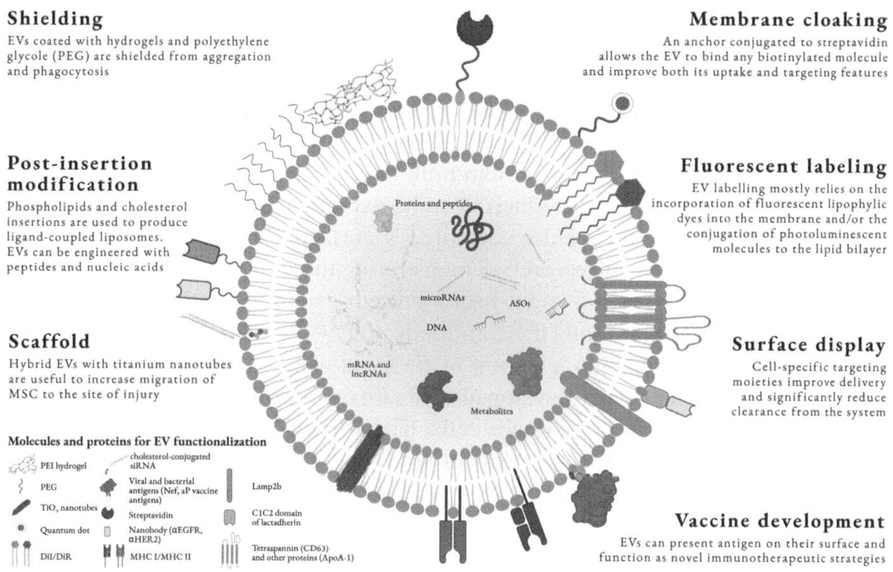

Fig. 6.1 Schematic representation of most employed methods for EV surface manipulations

modifications to the membrane with specific molecules while at the same time preserving the functionality and integrity of the vesicles (Fig. 6.1). There are two major approaches to surface engineering: either the EV-donor cell can be genetically or metabolically manipulated to produce EVs with membrane modification, or the surface of EVs can be physically or chemically modified after the EVs are isolated. Surface engineering is used for various purposes. The most common goal of EV surface engineering is to introduce moieties for a specific tissue targeting. Another purpose for surface modification is the attachment of fluorescent dyes to the exterior of EVs. To prevent EVs from uptake by non-target cells and to improve the overall pharmacokinetics, EVs can be shielded. Certain molecules can be difficult to load into EVs due to their size or polarity. Therefore, some surface engineering approaches focus on creating external vesicle cargo. Lastly, some EVs can trigger adverse effects due to antigens presented on their surface. This toxicity of antigenic EVs can be bypassed by modification of their surface epitopes. Some of the most recent examples of EV modification via surface engineering are described in the following sections.

2.1 *Surface Display*

Cell-specific targeting moieties (ligands) can be incorporated into the EV surface to improve the delivery to specific cells and tissues and significantly reduce the clearance of biologicals from the system. A more effective EV delivery can avoid or minimize the clearance of EVs and reduce the off-target effects. Therefore, a lower therapeutic dosage would be possible while maintaining the therapeutic efficacy.

Several approaches are being investigated for ligand display on EV surface. In one approach, targeting ligands are fused to known EV membrane proteins such as Lamp2b and CD63 [25, 37–39]. Another approach utilizes the C1C2 domain of the phosphatidylserine (PS) binding protein milk fat globule-EFG factor 8 (MFG-E8), also known as lactadherin. Lactadherin can be expressed from parent cells, can be added to purified EVs, and enables versatile EV surface display [40].

There is a large body of research demonstrating the use of lactadherin C1C2 to modify exosome surfaces and target fusion proteins, such as anti-EGFR nanobody, carcinoembryonic antigen, and HER2 antibody, to EVs [41–45]. A highly customizable platform for the functionalization of EVs via peptide and nanobody surface conjugation has been recently developed to obtain engineered EVs without the need for genetic manipulation of the donor cells. Cross-linking with a wide selection of antigens was tested on surface proteins of red blood cell–derived EVs using either sortase A or OaAEP1 bacterial protein-ligating enzymes. In particular, EV functionalization with α-EGFR, α-mCherry, and α-HER2 nanobodies allowed improved targeting of the correspondent antigen-expressing cells, while also favoring their uptake by H358 cells in a model for lung cancer. Successful treatment of the mouse model with these modified EVs loaded with paclitaxel further proved the clinical relevance of the system, inducing an increased apoptotic effect in the malignant cells compared to both the naked drug and the uncoated, loaded EVs [46].

Alternatively, when EV-producing cells are transfected with an expression plasmid, they package the plasmid DNA into EVs. If the plasmid encodes for a surface protein present in EVs, the EVs will simultaneously display the protein and carry the information to mediate its expression [47, 48]. EVs engineered this way have a similar nature to bacteriophages, making phage display screens possible. Komuro et al. utilized this approach to explore unique technique for EV-mediated targeted delivery. They engineered EVs with modified surface to deliver nucleic acid-based therapeutics. They used lactadherin (C1C2 domain) to display p88, a peptide with affinity for the human and mouse membrane small ion transport regulator (FXYD2) γa and demonstrated that β-cell targeting EVs can enrich plasmid DNA delivery to the murine pancreas, an inaccessible organ for both surgical and drug treatment in vivo [49].

2.2 Membrane Cloaking

An example of chemical modification of already generated EVs is the technology of cloaking. It can be used for labeling with fluorophores and also to enhance the targeting of EVs to the site of injury and uptake by specific target cells. An anchor conjugated to streptavidin is embedded directly into EV membrane surface ex vivo. Any biotinylated molecule can then be combined with tissue-targeted antibodies or homing peptides to facilitate EV uptake in target cells. Antes et al. successfully used this technology along with targeting EVs for enhanced uptake in cardiac fibroblasts, myoblasts, and ischemic myocardium implementing combinations of fluorescent tags, tissue targeting antibodies and homing peptide surface cloaks. Using a custom chemical synthesis service (NANOCS, Inc.), they generated a modular EV membrane-anchoring platform from conjugated streptavidin (STVDN) with 1,2-bis(dimethylphosphino)ethane:PEG 5k (DMPE-PEG) called DMPE-PEG-STVDN (abbreviated DPS). The DMPE-PEG fuses with the lipid bilayer of the EVs and provides an anchor for biotinylated molecules, such as ligands and fluorescent dyes [50, 51]. The DPS anchor was incubated with the biotinylated molecule (bio-FITC, or bio-PE, bio-antibody, bio-Homing peptide, bio-Qdot 655) and then the mixture was combined with multipotent, clonogenic cells of intrinsic cardiac origin—cardiosphere-derived cells (CDCs) or CDC-EVs. Similarly, the DPS anchor was incubated with α-DDR2 biotinylated rabbit polyclonal antibody, or rabbit IgG isotype control for cardiac fibroblast targeting. Biotinylated custom peptides were synthetized for muscle targeting and ischemia-targeting peptide. This proof-of-concept study presented three different types of cloaks (fluorescent molecules, targeting antibodies, and homing peptides) tested on various cell culture types, as well as animal models to verify tissue localization of engineered EVs.

This approach may be used to screen for targeting molecules to direct EVs to desired destinations in cell culture models in which EV uptake presents a challenge, and to direct EV delivery to organs and specific tissues in animal models [51].

2.3 *Post-insertion Approach*

Several studies have explored membrane post-insertion approach as a strategy for EV modification using methods based on phospholipid [52] and cholesterol-post-insertion [53–55]. In pharmaceutical field, this post-insertion approach is commonly used for production of ligand-coupled liposomes [56]. Using this method, EVs can be engineered with ligands as a whole antibody, antibody fragments, peptides, or other types of ligands. When applied to EV surface engineering, the ligands or choice are covalently linked to polyethylene glycol lipid micelles and then transferred in a simple incubation step from the micelles into the outer monolayer of pre-formed, drug-loaded EVs. This method offers a rapid and simple EV surface modification compared to traditional genetic engineering approaches. However further studies are necessary to fully determine the function of such EVs in vitro and in vivo. Zhang et al. have successfully used membrane post-insertion approach along with parental cell-assisted membrane modification strategy to develop magnetic and folate (FA)-modified microvesicles (MVs). They generated MVs modified with biotin and FA incorporated in their membrane. Then they conjugated the MVs with streptavidin-activated iron oxide nanoparticles (SA-IONPs) which allowed for efficient and rapid isolation of the vesicles from the supernatant using magnetic-activated sorting. When tested both in vitro and in vivo, the conjugated magnetic nanoparticles and FA conferred magnetic and ligand targeting activities on the vesicles, and when transformed into antitumor delivery platforms, the modified MVs exhibited significantly enhanced antitumor efficacy [57]. In another study researchers applied the post-insertion approach on both, mammalian cell–derived EVs and bacterial outer membrane vesicles (OMVs). They have optimized the technique for two types of mammalian cells and concluded that the post-insertion labeling efficiency depends on time, type of cells, and the method of culturing (suspension vs. adherent mode). By applying post-insertion approach to OMVs, they first showed that (possibly due to heavily glycosylated lipids present in the membrane) it does not modify the membrane of *Salmonella enterica* serovar typhimurium SL3261 strain, an engineered strain with hypervesiculation properties. The *Salmonella*-produced OMVs displayed rapid and more direct lipid insertion into their membrane. This may be due to a different lipid composition of the OMVs compared to the membrane from which they originate or the physicochemical characteristics of the OMVs. In conclusion, this study suggests different post-insertion mechanisms between mammalian EVs and bacterial OMVs. For mammalian EVs, cells incorporate the exogenous lipids into the cell membrane. The EVs shed from those modified cells are subsequently modified as well. As for bacteria, the exogenous lipids insert into OMVs directly, independent of the parental cells [58].

2.4 Shielding

In vivo, the delivery of EVs can also be impeded by interactions with and uptake by cells of the mononuclear phagocyte system (MPS) [59]. The MPS is a network of immune and architectural cells, located in organs such as the liver, spleen, and bone marrow, which remove foreign material from the bloodstream. A solution to this problem could be "shielding" or encapsulation of EVs in various materials. Suk et al. have shown that coating nanoparticles with hydrogels, mostly polyethylene glycol (PEG), shields them from aggregation and phagocytosis, and acts as sustained-release EV depot [60]. This treatment could also prolong the half-life of EVs [61]. Incorporation into biodegradable scaffolds such as polylactide (PLA) and polyethyleneimine (PEI) has also been successful [62]. Alternatively, 3D-printed ECM/gelatin-methacryloyl (GelMA)/EV scaffolds had promising results in regenerative application in bone and cartilage damage models [63]. A secondary approach consists of mimicking tumor immune evasion by displaying physiological markers of self on the EV surface. CD47, a member of the immunoglobulin superfamily, is the most important candidate to mediate a "don't eat me" signal on EVs through the inhibitory interaction with dendritic cells (DCs) and macrophages. Important implications for clinical applications depend on this feature. Direct correlation between the retention time in circulation of EVs and their levels of presenting CD47 antigen was demonstrated by Kamerkar et al. [64]. This was further confirmed by the observation of a decreased monocyte-mediated phagocytosis of the CD47-functionalized exosomes, an improved half-life for the encapsulated drug and better tissue distribution [65, 66]. Similarly, Rodrigues et al. have explored another promising approach to prevent unspecific phagocytic uptake. They created a synthetic, human-based self-peptide that specifically binds and signals to phagocytes to inhibit clearance of particles as small as liposomes and viruses, which enhanced delivery to the target cells [67].

2.5 *Scaffold*

Titanium is a widely used material in orthopedic surgery. Several different strategies have been explored to modify the titanium surface to enhance the performance and osseointegration of titanium implants for various applications. During the process of bone regeneration, the recruitment of an adequate number or progenitor cells or endogenous MSCs to the site of injury is very important for successful bone repair [68]. Zhao et al. fabricated EV hybrids with titanium nanotubes (TNT) and studied its effect on inflammation and osteogenesis. The EVs hybrid with TNT significantly increased osteogenic differentiation of human bone marrow–derived MSCs by increasing MSCs migration to the injury site. When they tested the effect of EVs TNT hybrid on inflammatory macrophages, there was a significant decrease in expression of inflammation-related genes, such as interleukin (IL)-6 and inducible nitric oxide synthase (iNOS), in the hybrid compared to the control [69].

2.6 *Vaccine Development*

Since EVs are able to deliver a wide range of cargo to target cells, it is possible to utilize EVs as a vaccination delivery mechanism. Also, unlike attenuated bacterial vaccines, EV-derived vaccines do not replicate and therefore are safer for clinical use. Raposo et al. have shown that human and murine B lymphocytes produce exosomes containing MHC II which can induce antigen-specific MHC class II-restricted T-cell responses [27]. This was the first study that suggested exosomes can physiologically present antigens, implying that exosomes could be utilized in immunotherapeutic strategies as vaccines, as well as biomarkers. Two years later, another study utilized exosomes as a novel cell-free vaccine. This study demonstrated that exosomes derived from peptide-pulsed dendritic cells (DEXs) presented functional MHC class I and II protein complexes and tumor antigens on their membrane and initiated tumor growth suppression in mice [70].

In the twenty-first century, more sophisticated techniques such as EV engineering have propelled the development of EVs as vaccines in areas other than oncology. A few years ago, the first results from clinical trials on EVs as vaccines were reported. Lamparski et al. have produced and characterized clinical-grade exosomes derived from antigen presenting cells. Their reproducible method and quality control assay produced an exosomal anticancer vaccine evaluated in several clinical trials [71–73]. Escudier et al. have produced DCs-derived exosomes capable of promoting T cell immune response and tumor rejection in stage III/IV melanoma patients [72]. Morse et al. tested safety and efficacy of autologous DCs-derived exosomes in the treatment of non-small cell lung cancer patients [73].

Currently, there are exosome-based vaccine candidates being developed for diseases such as hepatitis B, AIDS, cancer, coronavirus, and other pathogenic diseases. Promising approaches rely on the use of viral factors as the accessory proteins of retroviruses, current important pharmacologic targets to counteract human pathogens such as human immunodeficiency virus (HIV). These proteins fine-tune the host-virus interactions regulating key aspects of viral replication and pathogenicity of retroviruses. Among them, Nef protein plays an important role in HIV replication and onset of AIDS [74]. Nef was also discovered to accumulate in exosomes regardless of its activity [75]. Therefore, inactive Nef mutants were used to efficiently mediate exosomal incorporation of custom proteinaceous antigens fused to the C-terminal of the protein. Nef not only improves the loading but also elicits antigen-specific cytotoxic T Lymphocyte (CTL) activation against structural proteins of Ebola, influenza, and hepatitis B and C viruses [76–78]. Moreover, the strong immune response generated against recognized tumor antigens like HPV16-E7 and HER2/Neu has interesting potential for a new approach in cancer immunotherapy [79, 80].

Bacterial outer membrane vesicles (OMVs) are also gaining interest as a therapeutic platform against non-viral pathogens. There are pathogen-associated molecular patterns (PAMPs) present on the surface of bacteria, and these PAMPs are therefore also present on the surface of EVs shed from these bacteria. PAMPs are recognized by toll-like receptors (TLRs) and other pattern recognition receptors (PRRs) in infected cells and activate innate immune responses protecting the host from infection.

Vaccines based on OMVs have been licensed for the preventive treatment of *Neisseria meningitidis B* (Bexsero, VA-MENGOC-BC, Norway MenBVAC) and *Haemophilus influenzae type b I* (Hib) (PedvaxHib), as well as hexavalent formulation to protect against diphtheria, tetanus, pertussis, poliomyelitis, hepatitis B, and Hib (Vaxelis). An example of the additional benefits that engineered OMV-based vaccines offer is seen with the multi-antigen acellular pertussis vaccine (aP). Acellular pertussis vaccines (aP) have been more effective in reducing morbidity and mortality associated with pertussis compared to the traditional multi-antigen whole-cell pertussis vaccines (wP). However, due to the present-day circulation strains of *Bordetella pertussis*, the current aP vaccines do not provide durable immunity or adequate protection against the disease. A few years ago, scientists have designed and characterized a vaccine candidate based on OMVs. Using isogenic bacteria, they concluded that pertactin, a protein encoded by polymorphic gene present in the vaccine, is responsible for failure of the aP vaccines. Therefore, they designed a new multi-antigen aP vaccine formulation using outer membrane vesicles of *B. pertussis* in which antigens were presented in their native conformation and offered more effective protection to mice against lung infection than the current aP vaccines [81].

Many research are being conducted on OMVs from gram-negative bacteria, while the production of OMVs from gram-positive bacteria is poorly understood due to the thick peptidoglycan structure. Despite these challenges, a group of scientists engineered EVs by genetically modifying methicillin-resistant *Staphylococcus aureus* to overproduce detoxified cytolysins. The EVs were nontoxic and elicited immunogenic response in mice, triggering production of cytolysin-neutralizing antibodies and protecting the mice from lethal sepsis pointing out the possibility of EVs being used to alleviate the burden of antibiotic resistance [82].

Regarding the development of potential vaccine candidates, surface modification could increase the specificity of vaccines and improve the safety of EV-derived vaccines.

2.7 Fluorescent Labeling

One of the most common approaches to study the uptake of EVs is based on labelling the EV isolates with a fluorescent lipophilic membrane dye that is incorporated into the EV membrane. Several studies have implemented lipophilic dyes that incorporate themselves into the lipid bilayer membrane of EVs. Wiklander's group has used a near-infrared lipophilic dye DiR to label various murine and human cells [12]. Goes et al. have used DiI to label OMVs derived from myxobacteria Cystobacter velatus strain Cbv34 and Cystobacter ferrugineus strain Cbfe23 [83]. Other examples of lipophilic fluorescent dyes used for labeling of EVs are PKH26, [84] PKH67, [85] DiD, [86, 87] and CellMask [88]. Unfortunately, none these dyes are specific to the vesicular membrane as they are lipophilic molecules with broad affinity to lipids [89]. Despite the initial success of lipophilic dyes incorporation into EVs, there is a potential for complications—the dyes tend to aggregate into particles similar in size to EVs which are then difficult to separate from the labeled EVs. Fast photo-bleaching and false positive staining due to the incorporation of the dyes into other lipid-containing structures in addition to a long half-life of some dyes [90, 91] are all limiting factors to the use of this class of dyes.

Recently, Zhang et al. developed a method of labeling EVs through conjugation with high photoluminescent nanosized semiconductors, namely quantum dots (QDs), using click chemistry. They showed that QD-EV conjugates were stable in a biological environment and could be stored for up to a week. They were able to separate the QD-EV conjugates from unconjugated QDs and demonstrated that these can be suited for live and fixed high-resolution imaging of cells and tissue [92]. QDs for EV labeling have a great potential to reliably follow EVs in their role of drug and biologicals delivery systems for studies of EV localization and targeting.

3 Loading Cargo into EV

EVs can be successfully designed as delivery systems for different type of therapeutics (drugs, active molecules, nucleic acids, nanoparticles, etc.) for imaging, tracking, regeneration, and treatment of several diseases [93, 94].

Currently, various engineering strategies for an efficient cargo loading are available. Loading is classified as endogenous, when it is obtained through the modification of EV-secreting donor cells; otherwise, it is exogenous. The methods for exogenous loading of EVs first require the isolation and purification of the vesicles, and the subsequent internalization of the cargo of choice following different procedures to overcome the membrane barrier. Common techniques employed include passive loading by simple incubation with the required content [17], and active loading through physical treatments such as electroporation, sonication, extrusion, freeze-thaw cycles, and dialysis [95]. The appropriate loading method should be considered carefully according to the physical and chemical properties of the specific cargo. Notably, EV can encapsulate both hydrophilic and hydrophobic compounds, although loading of hydrophilic compounds (e.g., nucleic acid) proves to be more challenging. Efficient loading of hydrophobic molecules can be obtained by a simple co-incubating of the EV mixture with the drug. Conversely, hydrophilic compounds require the transient mechanical or chemical permeabilization of the hydrophobic lipid membrane to enter EVs. All these methodologies have different success rates, as some are more likely to promote aggregation of either the EVs or their cargo and therefore alter the physicochemical and morphological characteristics,thereby controlling their biological function [96].

3.1 Incubation of Therapeutics with EVs

One of the simplest and noninvasive ways to load small drug compounds into EV is to incubate the vesicles in a highly concentrated solution of the desired molecule, leading to its passive diffusion into the EVs following the concentration gradient [13]. This method has allowed the internalization of many hydrophobic drugs, such as paclitaxel (PTX), doxorubicin (DOX), and curcumin to enhance their therapeutic effect compared to free drugs. Curcumin-loaded EV has been the first drug internalized via co-incubation resulting in enhanced solubility, bioavailability, and stability [97]. This well-studied system revealed excellent performance and promising results in many areas of research, leading to clinical trials (https://clinicaltrials.gov/ct2/show/NCT01294072). The possibility to encapsulate FDA-approved anticancer drugs and nanoparticles into EV is of major interest to confer enhanced targeting specificity and lower cytotoxicity. Indeed, encapsulation of DOX in EVs can drastically reduce the severe side effects known to affect patients following chemotherapy. Recently, PTX packaged in embryonic stem cells–derived EVs showed enhanced therapeutic effect in glioblastoma [98]. Of great interest for both academic and industrial sectors is nucleic acid loading. As both hydrophilicity and high molecular weight are overcome, so is their intrinsic inability to cross the membrane barrier. Indeed, simple incubation with hydrophobic modified small interfering RNA (siRNA) was proven successful for EV loading [55].

Moreover, a new dialysis-dependent cargo loading system has been developed to enhance EV loading of negatively charged cargo such as siRNA, miRNA, and single-strand RNA, by creating a PH gradient across the membrane [99]. The strength of this passive loading strategy relies on a minimal manipulation alongside minimized destruction of the EV integrity, [100] which is critical to maintaining both integrity and stability of the nucleic acid to retain its functionality. Although this method has been successful for many compounds and EV sources, poor loading efficiency remains a major drawback. Moreover, loading quantitation is difficult to control due to physiochemical properties of both EV and cargo. The loading process is heavily affected by the concentration gradient of the molecules but it is concomitantly dependent on the lipophilic nature of the cargo and the pH of the solution, affecting the hydrophobic properties of the compounds [101]. It is worth considering the fact that the process of EV loading generates lipid aggregates and micelles regardless of the lipophilic or amphiphilic nature of the drug. This debris makes it difficult to separate from EVs, leading to some degree of contamination [18].

3.2 Electroporation

Electroporation is one of the most efficient, scalable, and GMP-compatible methods available for loading therapeutic molecules into EVs. This approach has been well utilized in several studies to encapsulate siRNA, miRNA, DNA, proteins, drugs, and nanoparticles in EVs [14, 15]. Electroporation is based on the application of an electrical field to an EV solution in the presence of the cargo of interest. The high voltage electrical charge creates transient nanometer-size pores on the EV phospholipidic membrane to enhance permeability, allowing the diffusion of the desired therapeutics. Several studies considered this method as a successful engineering strategy although with controversial loading efficiency records. On that note, Kooijmans et al. demonstrated that electroporation can cause siRNA precipitation, self-aggregation, and EV surface binding thus requiring co-purification with EVs. This pitfall might lead to a false interpretation of siRNA loading efficiency, which is not taken into consideration by most studies published so far [102]. Therefore, although electroporation is a straightforward and scalable GMP-engineering strategy, careful characterization of the real loading efficiency must be performed to avoid biases by nonencapsulated cargo. Washing and purification (e.g., density gradient) are the major steps to remove extracellular, protein-bound, and self-aggregate therapeutics before use the engineering EVs in functional assay. However, additional purification steps may disrupt EV structural integrity [20]. There is also a concern that electroporation could increase the size of EV by promoting the formation of aggregates. Such changes are not beneficial for maintaining the stability of

EV function in vivo and during storage. Therefore, the addition of a membrane stabilizer to the culture system, such as trehalose, may help to maintain the original size of EV and avoid EV fusion and aggregation [103]. Indeed, Johnsen et al. showed that adipose-derived stem cell (ASC) EV resulted in excessive EV aggregation after electroporation can be prevented using an optimized buffer containing the trehalose, without affecting the ASC-EV properties [104].

Although drugs can diffuse into EVs by simple co-incubation, several studies showed that drug loading efficiency can be significantly enhanced through electroporation. In a recent study by Gomari et al., electroporation technology was used to encapsulate DOX into EVs derived from bone marrow MSCs (BMSCs) for efficient targeting of a HER2-positive breast cancer model [105]. Furthermore, the generation of MUC1 aptamer-decorated MSC-EVs provided a selective guide of the loaded DOX through electroporation to the site of action improving the therapeutic index, while reducing its systemic toxicity [106]. A representative study generated hBMSC-derived-EV electroporated with KrasG12D siRNA therapeutic, which holds strong efficacy against pancreatic cancer [107]. The study has completed the recruitment for phase I clinical trial in September 2023 (https://clinicaltrials.gov/ct2/show/NCT03608631) and it is expected to publish its results during the second quarter of 2025. Of interest, Zhou et al. recently demonstrated a pancreatic duct adenocarcinoma (PDAC) targeting EV-based dual delivery biosystem (iEXO-OXA) for enhancing PDAC immunotherapy through induction of immunogenic-cell death (ICD) stimulus and immune suppression interfering [108]. EV isolated from BM-MSCs culture medium were electroporation-loaded with galectin-9-siRNA and superficially modified with oxaliplatin (OXA) prodrug as an ICD trigger. Such combined immunotherapy iEXO-AXO formulation preferentially delivered to the tumor site induced a synergistic effect enhancing the antitumor efficacy. Still further works on EV-derived stem cells show evidence that electroporation of LNA-antimiR-142-3p in BM-MSCs-derived EVs constitutes an efficient delivery system, improving breast cancer treatment [109].

Despite several studies and successes, loading efficiency and cargo aggregation remain dissatisfactory, and several optimization attempts to improve the protocols are still in progress [110, 111]. Nevertheless, it should be considered that several factors including electroporation devices, run setting, buffer composition, and the nature of EV and cargo being loading could contribute to the inconsistency of loading efficiency data across the scientific community.

3.3 Sonication

Sonication is an alternative method for loading hydrophilic compounds such as siRNA, miRNA, ssDNA, PTX, and DOX, inside EV. It is performed exposing the cargo mixture and EV solution to ultrasounds. High-frequency waves weaken the structure of the EV membrane and facilitate cargo loading while the membrane integrity is restored. This method has been shown to be effective in enhancing the uptake of the therapeutic target in multiple studies. In a study from Haney et al., PTX-loaded EV obtained by sonication possessed higher loading capability than the simple co-incubation [112]. Similarly, Kim et al. demonstrated that sonication has the highest loading efficiency of PTX into EV with strong antineoplastic effect compared to electroporation and co-incubation at room temperature [113]. Interestingly, honokiol, a newly discovered natural agent for cancer treatment, [114] has been successfully loaded into MSC-EV by the sonication method, resulting in a stronger antitumor effect compared to free drug [115]. Therefore, sonication is a useful method for loading certain drugs into EV as well as issues with nucleic acids, because of aggregation and/or denaturation [116]. Indeed, similar to electroporation, induced-membrane damage remains the main hurdle for large-scale applications. Moreover, sonication could alter the biophysical properties of MSC-EV (e.g., size and cellular uptake) compromising the preclinical and clinical outcome of their application. Therefore, optimization and standardization of procedure parameters deserve additional studies [16].

3.4 Freeze-Thaw Cycles, Extrusion, and Dialysis

Freeze-thaw treatment, extrusion, and dialysis enhance the EV cargo encapsulation during membrane recombination process upon its temporarily destabilization. EV loading by freeze-thaw cycle involves mixing EV and cargo at room temperature, followed by rapid freezing at −80 °C or liquid nitrogen, and subsequent thawing at room temperature; this is repeated three times to ensure proper encapsulation. This approach is simple and effective but affects the functionality and integrity of EVs, inducing major structural damage and aggregation. Moreover, the loading efficiency is typically lower than sonication and extrusion methods. To use extrusion strategy, EVs are mixed with the loading, and the resultant mixture is passed through a 100- to 140-nm pore-size membrane at room temperature. Hence, EV membrane collapses and blends with the cargo. However, EV surface structure could be affected as a result of membrane deformation impairing their immune privilege status. Moreover, changes in Z potential and cytotoxic effect have been observed. Recent reports demonstrated that desired therapeutic cargo could be encapsulated into EVs via hypotonic dialysis (HP). The HP procedure relies on osmosis; EVs and the cargo are mixed with hypotonic solution that causes the EV membrane permeabilization, allowing the cargo to pass through the membrane. Then, further dispersion of the mixture in an isotonic solution restores the integrity of the EVs and results in drug encapsulation. Drug-loaded EVs can be subsequently isolated by common purification strategies, such as size exclusion chromatography (SEC). Despite the higher drug loading efficiency (>11-fold) compared to passive co-incubation, protein degradation and EV surface characteristic changes may result in lower functionality [17]. In addition, dialysis method can be used to generate a pH gradient between the two sides of the EV plasma membrane improving miRNA and siRNA loading efficiently. Of note, since inconsistent results have been observed in several cellular uptake reports, detailed validation studies are warranted.

3.5 Endogenous Loading

Loading parental cells with the cargo of interest through co-incubation or cell genetic manipulation allows the production of EVs enriched with a desired therapeutic, exploiting the endogenous cellular cargo-sorting machinery. This method minimizes the impairment of the structures or of the biological activity of both cargoes and carriers. The cargo enriched in the donor cells is packed into EVs via endogenous sorting process during EV production; subsequently, therapeutic-loaded EVs were isolated and purified from culture medium.

The most common strategy for cell genome engineering is via direct transduction. Specific plasmids are transduced into EV-producing cells to induce the ectopic overexpression of the desired cargo and its spontaneous packaging into EVs. This method is mainly applied to load nucleic acid therapeutics (e.g., RNA, miRNA, siRNA) or protein drugs. Interestingly, it can be combined with novel approaches to increase the cargo active loading into the carrier. One strategy is sequence modification of the protein of interest, relying either on the addition of specific membrane targeting motifs (as protein-protein interaction) of proteins naturally enriched in EVs (e.g., Lamp2b, PDGFR, BASP1, tetraspanins, C1C2 domain) or by their entire fusion with the target protein [117]. To boost RNA loading into EVs, targeted and modular EV loading (TAMEL) methodology has been developed. It exploits the use of an engineered EV-enriched protein (EEP) able to tightly interact with nucleic acids. An example is the fusion of the RNA binding protein (RBP) MS2 with either Lamp2b or VSVG, which leads to the formation of EEPs able to enhance the RNA loading up to six-fold and 40-fold, respectively. Moreover, specific miRNA motif sequences have been identified as target site of RBP to sort them into EV, enabling the RNA EV loading manipulation [118]. Recently, Nguyen and Ferguson identified RNA sequences selectively loaded by cells into EV. By attaching those sequences to the desired RNA therapeutic, 100-fold loading increase was observed [119].

At present, an increasing number of research demonstrated the successful internalization of miRNA-based therapeutics in MSC-EVs by lentivirus-based transfection of the parental cells used in the treatment of malignant tumors, cardiovascular and neurological diseases. For example, miR-126, which plays a key role in promoting angiogenesis, was overexpressed in synovium MSCs to promote secretion of miR-126 rich EVs that enhanced angiogenesis compared to EV derived from unmodified parental cells [120]. Similarly, miR-100 delivered by MSC-EV was able to inhibit angiogenesis in breast cancer cells [121]. In another study, Luo et al. transfected MSCs with miR-122 expressing lentivirus and generated miR-122 enriched EVs with encouraging results in hepatocellular carcinoma treatment [122]. Intriguingly, the systemic administration of MSCs-derived EV carrying miR-379 upon transduction-triggered overexpression demonstrated a significantly reduced breast tumor growth unlike the administration of the engineering MSCs themselves that do not impact tumor growth [123]. In another breast cancer study, transfection of miR-222/223 inhibitors into MSCs promoted the secretion of EVs that were able to increase the sensitivity to carboplatin [124]. Interestingly, Baldari et al. constructed a lentiviral vector which expressed miR-125b linked with a specific ExoMotif sequence tag. By transfecting ASCs, miR-125b was efficiently loaded into derived EVs and reduced cell proliferation in hepatocellular carcinoma [125]. A similar study from Zhang et al. demonstrated the transduction of adipose tissue–derived MSCs to obtain miR-101-enriched EVs with tumor suppressive effect in osteosarcoma [126]. MiR-132 and miR-125b loaded in MSC-EVs demonstrated a protective cardiac function after myocardial ischemia and infarction respectively [127, 128]. Furthermore, the recombinant adenovirus-mediated miR-486 gene was used to modify the MSC, thus generating EV with high level of miRNA as a novel strategy for cardiac regeneration and repair [129]. A phase II clinical trial evaluated the effect of allogenic MSC-generated EV transfected by miR-124 (https://clinicaltrials.gov/ct2/show/NCT03384433) in patients affected by cerebrovascular accident (CVA).

In addition to delivering RNA therapeutics, stem cells can be transfected with gene encoding onto the targeted protein or peptide, resulting in protein or peptide carrying EVs. Until now, only a few studies have been carried out. Transduction-triggered overexpression of C-X-C chemokine receptor 4 (CXCR4) in MSCs, and the subsequent use of the harvested EVs, was found to promote angiogenesis in a rat model and prevent neonatal cardiomyocytes apoptosis in vitro [130]. Furthermore, MSCs have been engineered to overexpress bone morphogenetic protein 2 to generate EVs that contain high level of that protein with an enhanced capacity to promote bone regeneration [131]. Tumor necrosis factor-related apoptosis-inducing ligand (TRAIL) is a promising antitumor drug in preclinical research due to its ability to selectively induce cancer cells apoptosis. However, its benefit is limited, in part because of poor pharmacokinetics. Transduction of MSCs with lentiviruses expressing human TRAIL enables to obtain MSC-EVs delivering TRAIL with an effective antitumor effect [132, 133]. Of other interests, MSCs have been successfully manipulated by transfection to produce MSC-derived EVs carrying a cocktail of interfering RNAs as a novel therapeutic approach in COVID-19 management [134].

Although this endogenous method seems promising and simple, there are several factors to consider, including the possible side effect of the newly integrated molecules or transfection reagents on the viability and function of the parental cell, and the loading efficiency of the therapeutic molecule within the secreted EVs [135]. In addition, EVs derived from genetically modified cells would likely be categorized as advanced therapy medical products (ATMP), requiring more stringent safety testing compared to EVs derived from non-genetically modified stem cells, perhaps complicating clinical translation.

Another approach, simple and noninvasive, is directly incubating drugs or nanomaterials that can cross the cell membrane with parental cells, enabling the production of drug-enriched EVs. In this model, after the cells internalize the drug of interest, they are subjected to stress stimuli to increase the production of drug-loaded EVs, which can be collected and isolated from the conditioned culture media. The main interest of the spontaneous cell entry and/or EV exit strategy is its relatively easy and straightforward implementation. This approach can be exclusively employed to load lipophilic molecules and nanoparticles and has a limited loading capacity [18]. It was presented in a study by Pascucci et al. where MSCs were co-incubated with high dose of paclitaxel for 24 h to obtain MSCs-derived paclitaxel-loaded EVs with a strong therapeutic effect on pancreatic cancer cells [136]. Later, Kalimuthu et al. found that PTX-loaded MSC-EVs mimics were effective in inhibiting the breast tumor growth as compared to encapsulated MSC-EVs counterpart [137]. A similar approach was reported by Melzer et al. They effectively isolated PTX-loaded EV from hUCMSC pre-incubated with the drug, with an inhibitory effect on breast, lung and ovarian cancer growth and metastases [138]. Likewise, human gingival papilla mesenchymal stem cells (hGinPaMSCs) primed with a high dose of PTX-loaded EV were shown to promote anticancer activity [139]. Umbilical cord-MSC-EVs loaded with Vincristine showed an enhanced cytotoxicity compared both to the free drug and to unloaded EVs [140].

The same approach was used to incorporate hollow-gold nanoparticles functionalized with PEG into human placenta MSC-EVs allowing the optical hyperthermia for cancer therapy [141]. Incubation of iron oxide nanoparticles (Fe_2O_3) with MSCs resulted in the production of EV-loaded nanosystems that are taken up by naïve cells and disseminated throughout the body under the action of external magnetic navigation enabling the magnetic hyperthermia for tumor therapy [142]. Several studies reported that interference with the culture environment in which EVs are produced (e.g., hypoxia) affects the components secreted by EVs and their efficacy [143]. For example, it has been demonstrated that hypoxia treated MSCs-derived EV were superior to those produced under normoxic conditions in heart repair and regeneration [143–146]. Furthermore, introducing a nonactive biomolecules (e.g., growth factor, cytokines) into the cell culture medium increases the yield of EVs also modifying their content. For example, adipose tissue–derived MSCs cultured with PDGF (20 ng/mL) increase the EV production by 70%, enhancing the expression of pro-angiogenic factors while also inhibiting the anti-angiogenic factors [147].

Competing Interests The authors declare that they have no competing interests.

References

1. Liangsupree T, Multia E, Riekkola ML. Modern isolation and separation techniques for extracellular vesicles. J Chromatogr A. 2021;1636:461773. https://doi.org/10.1016/j.chroma.2020.461773.
2. Li J, He X, Deng Y, Yang C. An update on isolation methods for proteomic studies of extracellular vesicles in biofluids. Molecules. 2019;24. https://doi.org/10.3390/molecules24193516.
3. Thery C, et al. Minimal information for studies of extracellular vesicles 2018 (MISEV2018): a position statement of the International Society for Extracellular Vesicles and update of

the MISEV2014 guidelines. J Extracell Vesicles. 2018;7:1535750. https://doi.org/10.1080/20013078.2018.1535750.
4. Gardiner C, et al. Techniques used for the isolation and characterization of extracellular vesicles: results of a worldwide survey. J Extracell Vesicles. 2016;5:32945. https://doi.org/10.3402/jev.v5.32945.
5. Monguio-Tortajada M, Galvez-Monton C, Bayes-Genis A, Roura S, Borras FE. Extracellular vesicle isolation methods: rising impact of size-exclusion chromatography. Cell Mol Life Sci. 2019;76:2369–82. https://doi.org/10.1007/s00018-019-03071-y.
6. Benedikter BJ, et al. Ultrafiltration combined with size exclusion chromatography efficiently isolates extracellular vesicles from cell culture media for compositional and functional studies. Sci Rep. 2017;7:15297. https://doi.org/10.1038/s41598-017-15717-7.
7. Yan L, Wu X. Exosomes produced from 3D cultures of umbilical cord mesenchymal stem cells in a hollow-fiber bioreactor show improved osteochondral regeneration activity. Cell Biol Toxicol. 2020;36:165–78. https://doi.org/10.1007/s10565-019-09504-5.
8. Lener T, et al. Applying extracellular vesicles based therapeutics in clinical trials – an ISEV position paper. J Extracell Vesicles. 2015;4:30087. https://doi.org/10.3402/jev.v4.30087.
9. Bosch S, et al. Trehalose prevents aggregation of exosomes and cryodamage. Sci Rep. 2016;6:36162. https://doi.org/10.1038/srep36162.
10. Charoenviriyakul C, Takahashi Y, Nishikawa M, Takakura Y. Preservation of exosomes at room temperature using lyophilization. Int J Pharm. 2018;553:1–7. https://doi.org/10.1016/j.ijpharm.2018.10.032.
11. Kusuma GD, et al. To protect and to preserve: novel preservation strategies for extracellular vesicles. Front Pharmacol. 2018;9:1199. https://doi.org/10.3389/fphar.2018.01199.
12. Wiklander OP, et al. Extracellular vesicle in vivo biodistribution is determined by cell source, route of administration and targeting. J Extracell Vesicles. 2015;4:26316. https://doi.org/10.3402/jev.v4.26316.
13. Akuma P, Okagu OD, Udenigwe CC. Naturally occurring exosome vesicles as potential delivery vehicle for bioactive compounds. Front Sustain Food Syst. 2019;3. https://doi.org/10.3389/fsufs.2019.00023.
14. Zhang D, Lee H, Zhu Z, Minhas JK, Jin Y. Enrichment of selective miRNAs in exosomes and delivery of exosomal miRNAs in vitro and in vivo. Am J Physiol Lung Cell Mol Physiol. 2017;312:L110–21. https://doi.org/10.1152/ajplung.00423.2016.
15. Pomatto MAC, et al. Improved loading of plasma-derived extracellular vesicles to encapsulate antitumor miRNAs. Mol Ther Methods Clin Dev. 2019;13:133–44. https://doi.org/10.1016/j.omtm.2019.01.001.
16. Nizamudeen ZA, et al. Low-power sonication can alter extracellular vesicle size and properties. Cells. 2021;10. https://doi.org/10.3390/cells10092413.
17. Fuhrmann G, Serio A, Mazo M, Nair R, Stevens MM. Active loading into extracellular vesicles significantly improves the cellular uptake and photodynamic effect of porphyrins. J Control Release. 2015;205:35–44. https://doi.org/10.1016/j.jconrel.2014.11.029.
18. Piffoux M, et al. Engineering and loading therapeutic extracellular vesicles for clinical translation: a data reporting frame for comparability. Adv Drug Deliv Rev. 2021;178:113972. https://doi.org/10.1016/j.addr.2021.113972.
19. Busatto S, Iannotta D, Walker SA, Di Marzio L, Wolfram J. A simple and quick method for loading proteins in extracellular vesicles. Pharmaceuticals (Basel). 2021;14. https://doi.org/10.3390/ph14040356.
20. Baek G, Choi H, Kim Y, Lee HC, Choi C. Mesenchymal stem cell-derived extracellular vesicles as therapeutics and as a drug delivery platform. Stem Cells Transl Med. 2019;8:880–6. https://doi.org/10.1002/sctm.18-0226.
21. Yang J, Zhang X, Chen X, Wang L, Yang G. Exosome mediated delivery of miR-124 promotes neurogenesis after ischemia. Mol Ther Nucleic Acids. 2017;7:278–87. https://doi.org/10.1016/j.omtn.2017.04.010.

22. Abello J, Nguyen TDT, Marasini R, Aryal S, Weiss ML. Biodistribution of gadolinium- and near infrared-labeled human umbilical cord mesenchymal stromal cell-derived exosomes in tumor bearing mice. Theranostics. 2019;9:2325–45. https://doi.org/10.7150/thno.30030.
23. Moon GJ, et al. Application of mesenchymal stem cell-derived extracellular vesicles for stroke: biodistribution and MicroRNA study. Transl Stroke Res. 2019;10:509–21. https://doi.org/10.1007/s12975-018-0668-1.
24. Chen CC, et al. Elucidation of exosome migration across the blood-brain barrier model in vitro. Cell Mol Bioeng. 2016;9:509–29. https://doi.org/10.1007/s12195-016-0458-3.
25. Alvarez-Erviti L, et al. Delivery of siRNA to the mouse brain by systemic injection of targeted exosomes. Nat Biotechnol. 2011;29:341–5. https://doi.org/10.1038/nbt.1807.
26. Headland SE, et al. Neutrophil-derived microvesicles enter cartilage and protect the joint in inflammatory arthritis. Sci Transl Med. 2015;7:315ra190. https://doi.org/10.1126/scitranslmed.aac5608.
27. Raposo G, et al. B lymphocytes secrete antigen-presenting vesicles. J Exp Med. 1996;183:1161–72. https://doi.org/10.1084/jem.183.3.1161.
28. Valadi H, et al. Exosome-mediated transfer of mRNAs and microRNAs is a novel mechanism of genetic exchange between cells. Nat Cell Biol. 2007;9:654–9. https://doi.org/10.1038/ncb1596.
29. Thery C, Zitvogel L, Amigorena S. Exosomes: composition, biogenesis and function. Nat Rev Immunol. 2002;2:569–79. https://doi.org/10.1038/nri855.
30. Stoorvogel W, Kleijmeer MJ, Geuze HJ, Raposo G. The biogenesis and functions of exosomes. Traffic. 2002;3:321–30. https://doi.org/10.1034/j.1600-0854.2002.30502.x.
31. Kalluri R, LeBleu VS. The biology, function, and biomedical applications of exosomes. Science. 2020;367. https://doi.org/10.1126/science.aau6977.
32. Jurgielewicz BJ, Yao Y, Stice SL. Kinetics and specificity of HEK293T extracellular vesicle uptake using imaging flow cytometry. Nanoscale Res Lett. 2020;15:170. https://doi.org/10.1186/s11671-020-03399-6.
33. Mathieu M, Martin-Jaular L, Lavieu G, Thery C. Specificities of secretion and uptake of exosomes and other extracellular vesicles for cell-to-cell communication. Nat Cell Biol. 2019;21:9–17. https://doi.org/10.1038/s41556-018-0250-9.
34. Ozturk S, Elcin AE, Koca A, Elcin YM. Therapeutic applications of stem cells and extracellular vesicles in emergency care: futuristic perspectives. Stem Cell Rev Rep. 2021;17:390–410. https://doi.org/10.1007/s12015-020-10029-2.
35. Murphy DE, et al. Extracellular vesicle-based therapeutics: natural versus engineered targeting and trafficking. Exp Mol Med. 2019;51:1–12. https://doi.org/10.1038/s12276-019-0223-5.
36. Gudbergsson JM, Jonsson K, Simonsen JB, Johnsen KB. Systematic review of targeted extracellular vesicles for drug delivery – considerations on methodological and biological heterogeneity. J Control Release. 2019;306:108–20. https://doi.org/10.1016/j.jconrel.2019.06.006.
37. Reiner AT, et al. Concise review: developing best-practice models for the therapeutic use of extracellular vesicles. Stem Cells Transl Med. 2017;6:1730–9. https://doi.org/10.1002/sctm.17-0055.
38. Liang Y, Lehrich BM, Zheng S, Lu M. Emerging methods in biomarker identification for extracellular vesicle-based liquid biopsy. J Extracell Vesicles. 2021;10:e12090. https://doi.org/10.1002/jev2.12090.
39. Bellavia D, et al. Interleukin 3- receptor targeted exosomes inhibit in vitro and in vivo chronic myelogenous leukemia cell growth. Theranostics. 2017;7:1333–45. https://doi.org/10.7150/thno.17092.
40. Delcayre A, et al. Exosome display technology: applications to the development of new diagnostics and therapeutics. Blood Cells Mol Dis. 2005;35:158–68. https://doi.org/10.1016/j.bcmd.2005.07.003.
41. Wang JH, et al. Anti-HER2 scFv-directed extracellular vesicle-mediated mRNA-based gene delivery inhibits growth of HER2-positive human breast tumor xenografts by prodrug activation. Mol Cancer Ther. 2018;17:1133–42. https://doi.org/10.1158/1535-7163.MCT-17-0827.

42. Sedlik C, et al. Different immunogenicity but similar antitumor efficacy of two DNA vaccines coding for an antigen secreted in different membrane vesicle-associated forms. J Extracell Vesicles. 2014;3. https://doi.org/10.3402/jev.v3.24646.
43. Rountree RB, et al. Exosome targeting of tumor antigens expressed by cancer vaccines can improve antigen immunogenicity and therapeutic efficacy. Cancer Res. 2011;71:5235–44. https://doi.org/10.1158/0008-5472.CAN-10-4076.
44. Kooijmans SAA, Gitz-Francois J, Schiffelers RM, Vader P. Recombinant phosphatidylserine-binding nanobodies for targeting of extracellular vesicles to tumor cells: a plug-and-play approach. Nanoscale. 2018;10:2413–26. https://doi.org/10.1039/c7nr06966a.
45. Hartman ZC, et al. Increasing vaccine potency through exosome antigen targeting. Vaccine. 2011;29:9361–7. https://doi.org/10.1016/j.vaccine.2011.09.133.
46. Pham TC, et al. Covalent conjugation of extracellular vesicles with peptides and nanobodies for targeted therapeutic delivery. J Extracell Vesicles. 2021;10:e12057. https://doi.org/10.1002/jev2.12057.
47. Wan Y, et al. Nanoscale extracellular vesicle-derived DNA is superior to circulating cell-free DNA for mutation detection in early-stage non-small-cell lung cancer. Ann Oncol. 2018;29:2379–83. https://doi.org/10.1093/annonc/mdy458.
48. Kanada M, et al. Microvesicle-mediated delivery of minicircle DNA results in effective gene-directed enzyme prodrug cancer therapy. Mol Cancer Ther. 2019;18:2331–42. https://doi.org/10.1158/1535-7163.MCT-19-0299.
49. Komuro H, et al. Engineering extracellular vesicles to target pancreatic tissue in vivo. Nanotheranostics. 2021;5:378–90. https://doi.org/10.7150/ntno.54879.
50. Nag OK, Awasthi V. Surface engineering of liposomes for stealth behavior. Pharmaceutics. 2013;5:542–69. https://doi.org/10.3390/pharmaceutics5040542.
51. Antes TJ, et al. Targeting extracellular vesicles to injured tissue using membrane cloaking and surface display. J Nanobiotechnol. 2018;16:61. https://doi.org/10.1186/s12951-018-0388-4.
52. Kooijmans SAA, et al. PEGylated and targeted extracellular vesicles display enhanced cell specificity and circulation time. J Control Release. 2016;224:77–85. https://doi.org/10.1016/j.jconrel.2016.01.009.
53. Stremersch S, et al. Comparing exosome-like vesicles with liposomes for the functional cellular delivery of small RNAs. J Control Release. 2016;232:51–61. https://doi.org/10.1016/j.jconrel.2016.04.005.
54. O'Loughlin AJ, et al. Functional delivery of lipid-conjugated siRNA by extracellular vesicles. Mol Ther. 2017;25:1580–7. https://doi.org/10.1016/j.ymthe.2017.03.021.
55. Didiot MC, et al. Exosome-mediated delivery of hydrophobically modified siRNA for huntingtin mRNA silencing. Mol Ther. 2016;24:1836–47. https://doi.org/10.1038/mt.2016.126.
56. Moreira JN, Ishida T, Gaspar R, Allen TM. Use of the post-insertion technique to insert peptide ligands into pre-formed stealth liposomes with retention of binding activity and cytotoxicity. Pharm Res. 2002;19:265–9. https://doi.org/10.1023/a:1014434732752.
57. Zhang W, et al. Magnetic and folate functionalization enables rapid isolation and enhanced tumor-targeting of cell-derived microvesicles. ACS Nano. 2017;11:277–90. https://doi.org/10.1021/acsnano.6b05630.
58. Jiang L, et al. A bacterial extracellular vesicle-based intranasal vaccine against SARS-CoV-2 protects against disease and elicits neutralizing antibodies to wild-type and Delta variants. J Extracell Vesicles. 2022;11:e12192. https://doi.org/10.1002/jev2.12192.
59. Tsoi KM, et al. Mechanism of hard-nanomaterial clearance by the liver. Nat Mater. 2016;15:1212–21. https://doi.org/10.1038/nmat4718.
60. Suk JS, Xu Q, Kim N, Hanes J, Ensign LM. PEGylation as a strategy for improving nanoparticle-based drug and gene delivery. Adv Drug Deliv Rev. 2016;99:28–51. https://doi.org/10.1016/j.addr.2015.09.012.
61. Mardpour S, et al. Hydrogel-mediated sustained systemic delivery of mesenchymal stem cell-derived extracellular vesicles improves hepatic regeneration in chronic liver failure. ACS Appl Mater Interfaces. 2019;11:37421–33. https://doi.org/10.1021/acsami.9b10126.

62. Diomede F, et al. Three-dimensional printed PLA scaffold and human gingival stem cell-derived extracellular vesicles: a new tool for bone defect repair. Stem Cell Res Ther. 2018;9:104. https://doi.org/10.1186/s13287-018-0850-0.
63. Chen P, et al. Desktop-stereolithography 3D printing of a radially oriented extracellular matrix/mesenchymal stem cell exosome bioink for osteochondral defect regeneration. Theranostics. 2019;9:2439–59. https://doi.org/10.7150/thno.31017.
64. Kamerkar S, et al. Exosomes facilitate therapeutic targeting of oncogenic KRAS in pancreatic cancer. Nature. 2017;546:498–503. https://doi.org/10.1038/nature22341.
65. Belhadj Z, et al. A combined "eat me/don't eat me" strategy based on extracellular vesicles for anticancer nanomedicine. J Extracell Vesicles. 2020;9:1806444. https://doi.org/10.1080/20013078.2020.1806444.
66. Yang Z, et al. Large-scale generation of functional mRNA-encapsulating exosomes via cellular nanoporation. Nat Biomed Eng. 2020;4:69–83. https://doi.org/10.1038/s41551-019-0485-1.
67. Rodriguez PL, et al. Minimal "self" peptides that inhibit phagocytic clearance and enhance delivery of nanoparticles. Science. 2013;339:971–5. https://doi.org/10.1126/science.1229568.
68. Lin W, et al. Mesenchymal stem cells homing to improve bone healing. J Orthop Translat. 2017;9:19–27. https://doi.org/10.1016/j.jot.2017.03.002.
69. Zhao Q, et al. Surface engineering of titania nanotubes incorporated with double-layered extracellular vesicles to modulate inflammation and osteogenesis. Regen Biomater. 2021;8:rbab010. https://doi.org/10.1093/rb/rbab010.
70. Zitvogel L, et al. Eradication of established murine tumors using a novel cell-free vaccine: dendritic cell-derived exosomes. Nat Med. 1998;4:594–600. https://doi.org/10.1038/nm0598-594.
71. Lamparski HG, et al. Production and characterization of clinical grade exosomes derived from dendritic cells. J Immunol Methods. 2002;270:211–26. https://doi.org/10.1016/s0022-1759(02)00330-7.
72. Escudier B, et al. Vaccination of metastatic melanoma patients with autologous dendritic cell (DC) derived-exosomes: results of thefirst phase I clinical trial. J Transl Med. 2005;3:10. https://doi.org/10.1186/1479-5876-3-10.
73. Morse MA, et al. A phase I study of dexosome immunotherapy in patients with advanced non-small cell lung cancer. J Transl Med. 2005;3:9. https://doi.org/10.1186/1479-5876-3-9.
74. Basmaciogullari S, Pizzato M. The activity of Nef on HIV-1 infectivity. Front Microbiol. 2014;5:232. https://doi.org/10.3389/fmicb.2014.00232.
75. Lenassi M, et al. HIV Nef is secreted in exosomes and triggers apoptosis in bystander CD4+ T cells. Traffic. 2010;11:110–22. https://doi.org/10.1111/j.1600-0854.2009.01006.x.
76. Ferrantelli E, et al. Effluent and serum protein N-glycosylation is associated with inflammation and peritoneal membrane transport characteristics in peritoneal dialysis patients. Sci Rep. 2018;8:979. https://doi.org/10.1038/s41598-018-19147-x.
77. Chiozzini C, et al. N-terminal fatty acids of NEF(MUT) are required for the CD8(+) T-cell immunogenicity of in vivo engineered extracellular vesicles. Vaccines (Basel). 2020;8. https://doi.org/10.3390/vaccines8020243.
78. Anticoli S, et al. An exosome-based vaccine platform imparts cytotoxic T lymphocyte immunity against viral antigens. Biotechnol J. 2018;13:e1700443. https://doi.org/10.1002/biot.201700443.
79. Anticoli S, et al. Engineered exosomes emerging from muscle cells break immune tolerance to HER2 in transgenic mice and induce antigen-specific CTLs upon challenge by human dendritic cells. J Mol Med (Berl). 2018;96:211–21. https://doi.org/10.1007/s00109-017-1617-2.
80. Di Bonito P, et al. Antitumor HPV E7-specific CTL activity elicited by in vivo engineered exosomes produced through DNA inoculation. Int J Nanomedicine. 2017;12:4579–91. https://doi.org/10.2147/IJN.S131309.
81. Zurita ME, et al. A pertussis outer membrane vesicle-based vaccine induces lung-resident memory CD4 T cells and protection against Bordetella pertussis, including pertactin deficient strains. Front Cell Infect Microbiol. 2019;9:125. https://doi.org/10.3389/fcimb.2019.00125.

82. Wang X, Thompson CD, Weidenmaier C, Lee JC. Release of Staphylococcus aureus extracellular vesicles and their application as a vaccine platform. Nat Commun. 2018;9:1379. https://doi.org/10.1038/s41467-018-03847-z.
83. Goes A, et al. Myxobacteria-derived outer membrane vesicles: potential applicability against intracellular infections. Cells. 2020;9. https://doi.org/10.3390/cells9010194.
84. Puzar Dominkus P, et al. PKH26 labeling of extracellular vesicles: characterization and cellular internalization of contaminating PKH26 nanoparticles. Biochim Biophys Acta Biomembr. 2018;1860:1350–61. https://doi.org/10.1016/j.bbamem.2018.03.013.
85. Ekstrom K, et al. Monocyte exosomes stimulate the osteogenic gene expression of mesenchymal stem cells. PLoS One. 2013;8:e75227. https://doi.org/10.1371/journal.pone.0075227.
86. Tian T, et al. Exosome uptake through clathrin-mediated endocytosis and macropinocytosis and mediating miR-21 delivery. J Biol Chem. 2014;289:22258–67. https://doi.org/10.1074/jbc.M114.588046.
87. Tian T, Wang Y, Wang H, Zhu Z, Xiao Z. Visualizing of the cellular uptake and intracellular trafficking of exosomes by live-cell microscopy. J Cell Biochem. 2010;111:488–96. https://doi.org/10.1002/jcb.22733.
88. El-Andaloussi S, et al. Exosome-mediated delivery of siRNA in vitro and in vivo. Nat Protoc. 2012;7:2112–26. https://doi.org/10.1038/nprot.2012.131.
89. Lai CP, et al. Visualization and tracking of tumour extracellular vesicle delivery and RNA translation using multiplexed reporters. Nat Commun. 2015;6:7029. https://doi.org/10.1038/ncomms8029.
90. Takov K, Yellon DM, Davidson SM. Confounding factors in vesicle uptake studies using fluorescent lipophilic membrane dyes. J Extracell Vesicles. 2017;6:1388731. https://doi.org/10.1080/20013078.2017.1388731.
91. Botha J, Handberg A, Simonsen JB. Lipid-based strategies used to identify extracellular vesicles in flow cytometry can be confounded by lipoproteins: evaluations of annexin V, lactadherin, and detergent lysis. J Extracell Vesicles. 2022;11:e12200. https://doi.org/10.1002/jev2.12200.
92. Zhang M, Vojtech L, Ye Z, Hladik F, Nance E. Quantum dot labeling and visualization of extracellular vesicles. ACS Appl Nano Mater. 2020;3:7211–22. https://doi.org/10.1021/acsanm.0c01553.
93. Tian F, et al. Protein analysis of extracellular vesicles to monitor and predict therapeutic response in metastatic breast cancer. Nat Commun. 2021;12:2536. https://doi.org/10.1038/s41467-021-22913-7.
94. Mentkowski KI, Snitzer JD, Rusnak S, Lang JK. Therapeutic potential of engineered extracellular vesicles. AAPS J. 2018;20:50. https://doi.org/10.1208/s12248-018-0211-z.
95. Haney MJ, et al. Exosomes as drug delivery vehicles for Parkinson's disease therapy. J Control Release. 2015;207:18–30. https://doi.org/10.1016/j.jconrel.2015.03.033.
96. Sahoo S, et al. Therapeutic and diagnostic translation of extracellular vesicles in cardiovascular diseases: roadmap to the clinic. Circulation. 2021;143:1426–49. https://doi.org/10.1161/CIRCULATIONAHA.120.049254.
97. Sun D, et al. A novel nanoparticle drug delivery system: the anti-inflammatory activity of curcumin is enhanced when encapsulated in exosomes. Mol Ther. 2010;18:1606–14. https://doi.org/10.1038/mt.2010.105.
98. Zhu Q, et al. Embryonic stem cells-derived exosomes endowed with targeting properties as chemotherapeutics delivery vehicles for glioblastoma therapy. Adv Sci (Weinh). 2019;6:1801899. https://doi.org/10.1002/advs.201801899.
99. Jeyaram A, et al. Enhanced loading of functional miRNA cargo via pH gradient modification of extracellular vesicles. Mol Ther. 2020;28:975–85. https://doi.org/10.1016/j.ymthe.2019.12.007.
100. Kim H, et al. Engineered extracellular vesicles and their mimetics for clinical translation. Methods. 2020;177:80–94. https://doi.org/10.1016/j.ymeth.2019.10.005.

101. Haney MJ, et al. TPP1 delivery to lysosomes with extracellular vesicles and their enhanced brain distribution in the animal model of batten disease. Adv Healthc Mater. 2019;8:e1801271. https://doi.org/10.1002/adhm.201801271.
102. Kooijmans SAA, et al. Electroporation-induced siRNA precipitation obscures the efficiency of siRNA loading into extracellular vesicles. J Control Release. 2013;172:229–38. https://doi.org/10.1016/j.jconrel.2013.08.014.
103. Hood JL, Scott MJ, Wickline SA. Maximizing exosome colloidal stability following electroporation. Anal Biochem. 2014;448:41–9. https://doi.org/10.1016/j.ab.2013.12.001.
104. Johnsen KB, et al. Evaluation of electroporation-induced adverse effects on adipose-derived stem cell exosomes. Cytotechnology. 2016;68:2125–38. https://doi.org/10.1007/s10616-016-9952-7.
105. Gomari H, Forouzandeh Moghadam M, Soleimani M, Ghavami M, Khodashenas S. Targeted delivery of doxorubicin to HER2 positive tumor models. Int J Nanomedicine. 2019;14:5679–90. https://doi.org/10.2147/IJN.S210731.
106. Bagheri E, et al. Targeted doxorubicin-loaded mesenchymal stem cells-derived exosomes as a versatile platform for fighting against colorectal cancer. Life Sci. 2020;261:118369. https://doi.org/10.1016/j.lfs.2020.118369.
107. Mendt M, et al. Generation and testing of clinical-grade exosomes for pancreatic cancer. JCI Insight. 2018;3. https://doi.org/10.1172/jci.insight.99263.
108. Zhou W, et al. Pancreatic cancer-targeting exosomes for enhancing immunotherapy and reprogramming tumor microenvironment. Biomaterials. 2021;268:120546. https://doi.org/10.1016/j.biomaterials.2020.120546.
109. Naseri Z, Oskuee RK, Forouzandeh-Moghadam M, Jaafari MR. Delivery of LNA-antimiR-142-3p by mesenchymal stem cells-derived exosomes to breast cancer stem cells reduces tumorigenicity. Stem Cell Rev Rep. 2020;16:541–56. https://doi.org/10.1007/s12015-019-09944-w.
110. Reshke R, et al. Reduction of the therapeutic dose of silencing RNA by packaging it in extracellular vesicles via a pre-microRNA backbone. Nat Biomed Eng. 2020;4:52–68. https://doi.org/10.1038/s41551-019-0502-4.
111. Lennaard AJ, Mamand DR, Wiklander RJ, El Andaloussi S, Wiklander OPB. Optimised electroporation for loading of extracellular vesicles with doxorubicin. Pharmaceutics. 2021;14. https://doi.org/10.3390/pharmaceutics14010038.
112. Haney MJ, et al. Macrophage-derived extracellular vesicles as drug delivery systems for triple negative breast cancer (TNBC) therapy. J Neuroimmune Pharmacol. 2020;15:487–500. https://doi.org/10.1007/s11481-019-09884-9.
113. Kim S, et al. Extracellular vesicles with high dual drug loading for safe and efficient combination chemo-phototherapy. Biomater Sci. 2022;10:2817. https://doi.org/10.1039/d1bm02005f.
114. Kanchanapally R, et al. Exosomal formulation escalates cellular uptake of honokiol leading to the enhancement of its antitumor efficacy. ACS Omega. 2020;5:23299–307. https://doi.org/10.1021/acsomega.0c03136.
115. Sun Z, et al. Roles of mesenchymal stem cell-derived exosomes in cancer development and targeted therapy. Stem Cells Int. 2021;2021:9962194. https://doi.org/10.1155/2021/9962194.
116. Lamichhane TN, et al. Oncogene knockdown via active loading of small RNAs into extracellular vesicles by sonication. Cell Mol Bioeng. 2016;9:315–24. https://doi.org/10.1007/s12195-016-0457-4.
117. Dooley K, et al. A versatile platform for generating engineered extracellular vesicles with defined therapeutic properties. Mol Ther. 2021;29:1729–43. https://doi.org/10.1016/j.ymthe.2021.01.020.
118. Nasiri Kenari A, Cheng L, Hill AF. Methods for loading therapeutics into extracellular vesicles and generating extracellular vesicles mimetic-nanovesicles. Methods. 2020;177:103–13. https://doi.org/10.1016/j.ymeth.2020.01.001.
119. Ferguson S, Kim S, Lee C, Deci M, Nguyen J. The phenotypic effects of exosomes secreted from distinct cellular sources: a comparative study based on miRNA composition. AAPS J. 2018;20:67. https://doi.org/10.1208/s12248-018-0227-4.

120. Tao SC, et al. Chitosan wound dressings incorporating exosomes derived from MicroRNA-126-overexpressing synovium mesenchymal stem cells provide sustained release of exosomes and heal full-thickness skin defects in a diabetic rat model. Stem Cells Transl Med. 2017;6:736–47. https://doi.org/10.5966/sctm.2016-0275.
121. Pakravan K, et al. MicroRNA-100 shuttled by mesenchymal stem cell-derived exosomes suppresses in vitro angiogenesis through modulating the mTOR/HIF-1alpha/VEGF signaling axis in breast cancer cells. Cell Oncol (Dordr). 2017;40:457–70. https://doi.org/10.1007/s13402-017-0335-7.
122. Lou G, et al. Exosomes derived from miR-122-modified adipose tissue-derived MSCs increase chemosensitivity of hepatocellular carcinoma. J Hematol Oncol. 2015;8:122. https://doi.org/10.1186/s13045-015-0220-7.
123. O'Brien KP, et al. Employing mesenchymal stem cells to support tumor-targeted delivery of extracellular vesicle (EV)-encapsulated microRNA-379. Oncogene. 2018;37:2137–49. https://doi.org/10.1038/s41388-017-0116-9.
124. Bliss SA, et al. Mesenchymal stem cell-derived exosomes stimulate cycling quiescence and early breast cancer dormancy in bone marrow. Cancer Res. 2016;76:5832–44. https://doi.org/10.1158/0008-5472.CAN-16-1092.
125. Baldari S, Di Rocco G, Magenta A, Picozza M, Toietta G. Extracellular vesicles-encapsulated MicroRNA-125b produced in genetically modified mesenchymal stromal cells inhibits hepatocellular carcinoma cell proliferation. Cells. 2019;8. https://doi.org/10.3390/cells8121560.
126. Zhang K, et al. Extracellular vesicle-mediated delivery of miR-101 inhibits lung metastasis in osteosarcoma. Theranostics. 2020;10:411–25. https://doi.org/10.7150/thno.33482.
127. Xiao C, et al. Transplanted mesenchymal stem cells reduce autophagic flux in infarcted hearts via the exosomal transfer of miR-125b. Circ Res. 2018;123:564–78. https://doi.org/10.1161/CIRCRESAHA.118.312758.
128. Ma T, et al. MicroRNA-132, delivered by mesenchymal stem cell-derived exosomes, promote angiogenesis in myocardial infarction. Stem Cells Int. 2018;2018:3290372. https://doi.org/10.1155/2018/3290372.
129. Fang L, et al. Effects of exosomes derived from miR-486 gene modified umbilical cord mesenchymal stem cells on biological characteristics of rat Cardiomyocytes. Zhongguo Shi Yan Xue Ye Xue Za Zhi. 2018;26:1531–7. https://doi.org/10.7534/j.issn.1009-2137.2018.05.045.
130. Liu L, et al. Exosomes derived from mesenchymal stem cells rescue myocardial Ischaemia/reperfusion injury by inducing cardiomyocyte autophagy via AMPK and Akt pathways. Cell Physiol Biochem. 2017;43:52–68. https://doi.org/10.1159/000480317.
131. Huang CC, et al. Functionally engineered extracellular vesicles improve bone regeneration. Acta Biomater. 2020;109:182–94. https://doi.org/10.1016/j.actbio.2020.04.017.
132. Shamili FH, et al. Exosomes derived from TRAIL-engineered mesenchymal stem cells with effective anti-tumor activity in a mouse melanoma model. Int J Pharm. 2018;549:218–29. https://doi.org/10.1016/j.ijpharm.2018.07.067.
133. Yuan Z, Kolluri KK, Gowers KH, Janes SM. TRAIL delivery by MSC-derived extracellular vesicles is an effective anticancer therapy. J Extracell Vesicles. 2017;6:1265291. https://doi.org/10.1080/20013078.2017.1265291.
134. Jamalkhah M, et al. MSC-derived exosomes carrying a cocktail of exogenous interfering RNAs an unprecedented therapy in era of COVID-19 outbreak. J Transl Med. 2021;19:164. https://doi.org/10.1186/s12967-021-02840-3.
135. Gomzikova MO, James V, Rizvanov AA. Therapeutic application of mesenchymal stem cells derived extracellular vesicles for immunomodulation. Front Immunol. 2019;10:2663. https://doi.org/10.3389/fimmu.2019.02663.
136. Pascucci L, et al. Paclitaxel is incorporated by mesenchymal stromal cells and released in exosomes that inhibit in vitro tumor growth: a new approach for drug delivery. J Control Release. 2014;192:262–70. https://doi.org/10.1016/j.jconrel.2014.07.042.
137. Kalimuthu S, et al. A new approach for loading anticancer drugs into mesenchymal stem cell-derived exosome mimetics for cancer therapy. Front Pharmacol. 2018;9:1116. https://doi.org/10.3389/fphar.2018.01116.

138. Melzer C, et al. Taxol-loaded MSC-derived exosomes provide a therapeutic vehicle to target metastatic breast cancer and other carcinoma cells. Cancers (Basel). 2019;11. https://doi.org/10.3390/cancers11060798.
139. Cocce V, et al. In vitro anticancer activity of extracellular vesicles (EVs) secreted by gingival mesenchymal stromal cells primed with paclitaxel. Pharmaceutics. 2019;11. https://doi.org/10.3390/pharmaceutics11020061.
140. Del Fattore A, et al. Differential effects of extracellular vesicles secreted by mesenchymal stem cells from different sources on glioblastoma cells. Expert Opin Biol Ther. 2015;15:495–504. https://doi.org/10.1517/14712598.2015.997706.
141. Sancho-Albero M, et al. Cancer-derived exosomes loaded with ultrathin palladium nanosheets for targeted bioorthogonal catalysis. Nat Catal. 2019;2:864–72. https://doi.org/10.1038/s41929-019-0333-4.
142. Altanerova U, et al. Human mesenchymal stem cell-derived iron oxide exosomes allow targeted ablation of tumor cells via magnetic hyperthermia. Int J Nanomedicine. 2017;12:7923–36. https://doi.org/10.2147/IJN.S145096.
143. Zhu LP, et al. Hypoxia-elicited mesenchymal stem cell-derived exosomes facilitates cardiac repair through miR-125b-mediated prevention of cell death in myocardial infarction. Theranostics. 2018;8:6163–77. https://doi.org/10.7150/thno.28021.
144. Zhu J, et al. Myocardial reparative functions of exosomes from mesenchymal stem cells are enhanced by hypoxia treatment of the cells via transferring microRNA-210 in an nSMase2-dependent way. Artif Cells Nanomed Biotechnol. 2018;46:1659–70. https://doi.org/10.1080/21691401.2017.1388249.
145. Hu J, et al. MiR-215 is induced post-transcriptionally via HIF-Drosha complex and mediates glioma-initiating cell adaptation to hypoxia by targeting KDM1B. Cancer Cell. 2016;29:49–60. https://doi.org/10.1016/j.ccell.2015.12.005.
146. Bian S, et al. Extracellular vesicles derived from human bone marrow mesenchymal stem cells promote angiogenesis in a rat myocardial infarction model. J Mol Med (Berl). 2014;92:387–97. https://doi.org/10.1007/s00109-013-1110-5.
147. Lopatina T, et al. Platelet-derived growth factor regulates the secretion of extracellular vesicles by adipose mesenchymal stem cells and enhances their angiogenic potential. Cell Commun Signal. 2014;12:26. https://doi.org/10.1186/1478-811X-12-26.

Chapter 7
Mesenchymal Stem Cells–Derived Extracellular Vesicles in Therapeutic Applications

Adele Tanzi, Roberta Verta, and Benedetta Bussolati

Abbreviations

EVs	extracellular vesicles
MSCs	mesenchymal stem/stromal cells
MSC-EVs	mesenchymal stem/stromal cells-derived EVs
sEVs	small extracellular vesicles
BM-MSCs	bone marrow-derived mesenchymal stem/stromal cells
AT-MSCs	adipose tissue-derived mesenchymal stem/stromal cells
UC-MSCs	umbilical cord-derived mesenchymal stem/stromal cells
UCB-MSCs	umbilical cord blood-derived mesenchymal stem/stromal cells
P-MSCs	placenta-derived mesenchymal stem/stromal cells.
COVID-19	coronavirus disease
SARS-CoV-2	severe acute respiratory syndrome- coronavirus-2
ARDS	acute respiratory distress syndrome

1 MSCs and MSC-EVs

Mesenchymal stem/stromal cells (MSCs) are stromal cells that possess the capacity to self-renew and exhibit multilineage differentiation [3]. Due to their properties, including their capacity to migrate to the injured sites, MSCs have revealed great potential in regenerative medicine since the 1970s [3]. In 2006, the International

A. Tanzi · B. Bussolati (✉)
Department of Medical Science, University of Turin, Turin, Italy
e-mail: adele.tanzi@unito.it; benedetta.bussolati@unito.it

R. Verta
Department of Molecular Biotechnology and Health Sciences, University of Turin, Turin, Italy
e-mail: roberta.verta@unito.it

P. V. Pham, S. Fagoonee (eds.), *Stem Cell-Derived Exosomes and Microvesicles*, Stem Cells in Clinical Applications, https://doi.org/10.1007/978-3-031-90698-5_7

Society for Cell and Gene Therapy (ISCT) established the minimal defining criteria of human MSCs, which state as follows: "First, MSC must be plastic-adherent when maintained in standard culture conditions. Second, MSC must express CD105, CD73 and CD90, and lack expression of CD45, CD34, CD14 or CD11b, CD79alpha or CD19 and HLA-DR surface molecules. Third, MSC must differentiate to osteoblasts, adipocytes and chondroblasts in vitro" [19].

In humans, MSCs have been isolated from many different sources: bone marrow (BM), umbilical cord (UC), adipose tissue (AT), umbilical cord blood (UCB), synovial fluid, dental pulp, endometrial tissues, and others [66]. Nonetheless, only a few of them are clinically useful, with bone marrow being the most enriched source of MSCs. However, the isolation from this source is very complicated and invasive [53]. Umbilical cord blood (UCB) is a more easily accessible source of MSCs and considered to be safe, noninvasive, and devoid of possible senescence and inflammatory conditions of the donor, thus displaying excellent therapeutic characteristics [35, 54, 66].

MSCs were initially called "mesenchymal stem cells" and thought to act therapeutically as stem cells via cellular differentiation and cell replacement [15]. However, several studies showed that their effects are mediated mainly by paracrine factors, therefore challenging their stem cell nature [27, 80].

For this reason, they may be more appropriately named "mesenchymal stromal cells" [15, 29]. Furthermore, increasing evidence suggests that MSCs exert most of their paracrine effects through the release of extracellular vesicles (EVs) [94]. For this reason, nowadays, researchers are studying possible therapeutic approaches directly based on MSC-derived EVs (MSC-EVs) administration. A similar approach would offer many advantages if compared to an MSC-based therapy: low toxicity, low immunogenicity and tumorigenicity, and easier manipulation. Another important advantage is that treatment with MSC-EVs is not permanent and thus, in case of any adverse effect, can be interrupted more easily than cells [6]. Moreover, the intravenous administration of MSCs may cause aggregation that could lead to embolism, while EVs possess a higher stability in circulation [83, 84].

2 MSC-Small EVs (MSC-sEVs)

As mentioned in the previous chapters, EVs are released from cells as a heterogeneous population which can be further classified into three fractions: small EVs, large EVs, and apoptotic bodies [93]. Recently, researchers fractionated MSC-EVs in medium/large EVs (100–1000 nm) using a 10,000 × g ultracentrifugation (10 k fraction) and in small MSC EVs (<100 nm) using a subsequent 100,000 × g ultracentrifugation (100 k fraction), with the aim of identifying the potentially most relevant MSC-EVs subpopulation [13, 74]. In vitro and in vivo preclinical experiments clearly showed that the 100 k fraction was the main fraction responsible for functional and morphological tissue regeneration [13, 74]. Indeed, the two fractions appeared biochemically and functionally distinct [13, 96], with the 100 k fraction (small MSC-EVs) considered nowadays the proactive subpopulation retaining the therapeutic activity [94].

EVs can also be categorized into exosomes and microvesicles depending on their biogenesis from multivesicular bodies or cell surface budding, respectively [81]. Exosomes are commonly thought to be smaller than microvesicles, while actually the two EV types cannot be differentiated only by size, since their respective size ranges overlap. Therefore, classifying sEVs as "exosomes" would be incorrect, since the evidence of an endosomal biogenesis pathway is required to define exosomes. Moreover, the preparation of a pure MSC-exosome population, by depleting non-endosome-origin vesicles, would be experimentally challenging and of academic rather than clinical interest [81]. As a consequence, even if providing little information on the biology of EVs, the term "sEV" is recommended, since it is independent of the site of subcellular origin and defines an EV population that could be universally prepared [81, 90]. Moreover, as we mentioned, MSC-sEVs refer to the MSC-EVs fraction for which the therapeutic activity has been demonstrated [94]. Nevertheless, the scientific community is still widely using the term "exosomes."

3 Preclinical Trials with MSC-sEVs

During the past years, MSC-EVs have shown encouraging therapeutic effects in preclinical (in vitro and in vivo) models of various pathologies. For instance, in case of liver injuries, studies demonstrate the ability of MSC-EVs to induce hepatocyte regeneration and to inhibit fibrosis and apoptosis [31, 50, 78]. Concerning cardiovascular diseases, MSC-EV treatment has been shown to reduce infarct size, enhance tissue repair, and increase angiogenesis in animal models, therefore exerting a cardioprotective effect [4, 8, 43]. MSC-EV anti-apoptotic activities were also effective in promoting kidney regeneration in severe combined immunodeficiency (SCID) mice with glycerol-induced acute kidney injury [11]. Later works confirmed the protective properties of MSC-EVs in both acute and chronic renal damage, with a reduction of fibrosis, apoptosis, and tubular atrophy [12, 22, 28, 108]. Concerning lung injury, various studies have shown that MSC-EV treatments reduce lung edema and inflammation, while improving pulmonary hypertension, ventricular hypertrophy, and lung vascular remodeling [45, 46, 109]. MSC-EVs treatments have shown therapeutic potential also in preclinical models of various neurodegenerative disorders. In a Parkinson's disease rat model, for instance, MSC-EVs treatment significantly helped in improving functional recovery and angio-neurogenesis, reducing neuroinflammation, thus providing a neuroprotective effect. Moreover, MSC-exosomes had strong therapeutic potential in treating Alzheimer's disease [79]. It was also demonstrated that MSC-exosomes can significantly increase wound healing in a skin-deep second-degree burn wound in a rat model [103]. Finally, in the literature, there are also examples of bone, muscle, and cartilage regeneration induced by MSC-EVs treatment [83, 84].

4 Clinical Trials on MSC-sEVs Effects in Acute and Chronic Organ Dysfunction

Besides the numerous preclinical studies, various clinical trials have been conducted testing MSC-sEV effects in acute and chronic pathologies (Table 7.1). In 2014, the first-in-human EV application was reported in a study conducted on graft-versus-host disease (GvHD) [38]. An exosome-enriched fraction processed from collected bone marrow-MSC supernatants, referred to as "MSC1 Exo", was used as treatment of one therapy-refractory patient with GvHD. Before the MSC1 Exo administration, to reduce the risk of any unexpected reaction of the patient's immune response to the treatment, the authors tested the immune-modulatory impacts on peripheral blood mononucleated cells and natural killer cells in mixed lymphocyte reactions. After demonstrating the immunosuppressive activities of MSC1 Exo, the patient was treated. Initially, to reduce the risk of potential side effects, only a tenth of MSC-sEVs was administered. No side effects were observed within 2 days, thus unit amounts were progressively increased and administered every 2–3 days until a four-time dosage was reached. The authors reported that all applications were well tolerated, and no side effects were detected. Concerning the efficacy, a significant improvement in the clinical GvHD symptoms was observed shortly after the start of the MSC1 Exo therapy. The patient remained stable for several months, but died of pneumonia 7 months after MSC1 Exo application. However, the authors concluded that MSC-derived exosomes may be a potentially safe tool for treating therapy-refractory GvHD and other inflammation-associated diseases [38].

Table 7.1 List of completed clinical trials on the use of EVs from different MSC sources in various acute and chronic pathologies

References	Disease	EV source	Administration route
Kordelas et al. [38]	Graft versus host disease	BM-MSCs	Not reported
Nassar et al. [60]	Chronic kidney disease	UCB-MSCs	Intravenous infusion and intra-arterial injection
Zhang et al. [106]	Macular holes	UC-MSCs	Intravitreal injection
Warnecke et al. [89]	Hearing loss caused by cochlear implant procedures	UC-MSCs	Intracochlear injections
Zhou et al. [107]	Graft versus host disease—Associated dry eye disease	UC-MSCs	Eye drops
Nazari et al. [61]	Perianal fistula (Crohn's disease associated)	UC-MSCs	Injection along the fistula tract
Pak et al. [62]	Perianal fistula (non-Crohn's disease associated)	P-MSCs	Injection along the fistula tract

Abbreviations: *BM-MSCs* bone marrow–derived mesenchymal stem/stromal cells, *UC-MSCs* umbilical cord–derived mesenchymal stem/stromal cells, *UCB-MSCs* umbilical cord blood–derived mesenchymal stem/stromal cells, *P-MSCs* placenta-derived mesenchymal stem/stromal cells. The clinical trials research data is obtained from trialsearch.who.int, clinicaltrials.gov, and clinicaltrialsregister.eu

In 2016, the first clinical trial on MSC-sEVs in chronic organ dysfunction was published. This study consisted of a single-center, randomized, placebo-controlled, phase II/III clinical pilot study conducted on chronic kidney disease (CKD) patients who were treated with umbilical cord blood-MSC-EVs [60]. Forty CKD patients, stage III and IV, were recruited and separated into two groups: half patients as treatment group and the other half as a matching placebo group. Two doses of cell-free cord blood-MSC-EVs at 100 μg/kg/dose had been administered to patients of the treatment group. In particular, the first dose was an intravenous injection, while the second, with 1 week apart, was intra-arterial injection targeting the kidney. Renal function parameters such as blood urea, serum creatinine, urinary albumin-to-creatinine ratio, and estimated glomerular filtration rate (eGFR) were analyzed to assess kidney functions. In addition, TNF-α, TGF-β1, and IL-10 were used to evaluate the amelioration of the inflammatory immune activity. Participants of the treatment group showed significant improvement of eGFR, serum creatinine level, blood urea, and albumin-to-creatinine ratio. Moreover, they also exhibited a significant increase in plasma levels of TGF-β1 and IL-10 and a significant decrease in plasma levels of TNF-α. At variance, participants of the control group did not show significant improvement in any of the mentioned parameters. This clinical trial has proved the safety of the described treatment, as well as its positive effect on the inflammatory immune reaction and the improvement of the overall kidney function in grade III-IV CKD patients [60].

Recently, another study reported the first-in-human intracochlear application of human MSC-EVs. In this case, MSC-EV treatment is used to reduce inflammation associated with cochlear implantation, and so to limit the potential damage to the inner ear from a foreign body reaction and inflammation induced by the opening of the cochlea and by the insertion of the electrode array [89]. Umbilical cord tissue-derived EVs have been chosen for the treatment. The patient (male; age at implantation: 55 years) received intracochlear UC-MSC-EV injection during the cochlear implantation surgical process. The administration occurred through a syringe connected to an inner ear catheter. The total delivery volume was estimated to range from 20 to 40 μL corresponding to a dose of 2×10^9 to 4×10^9 UC-MSC-EVs in total. After administration, no signs of acute systemic or local toxicity were noted within an observation period of 5 days. Most importantly, the patient showed an initial improvement in speech intelligibility, which stabilized in the 24-month follow-up period. These observations suggested that the local administration of allogeneic UC-MSC-EVs to the inner ear is safe in principle. However, the experience from one single patient treatment is a limiting factor and is not sufficient; therefore, a clinical phase 1 trial is being planned [89].

Another clinical trial investigated whether UC-MSCs-derived EVs could alleviate dry eye symptoms in patients with chronic GvHD. The UC-MSC-exosomes treatment in this case was local, administered as eye drops, four times a day for 14 days. The authors observed a significantly reduced inflammation and an amelioration of epithelial recovery in patients treated with MSC-EVs compared to placebo. Moreover, miRNA-204 was identified as a possible crucial player in the observed pro-regenerative effect [107].

In Tehran, Iran, a phase 1 clinical trial was conducted to evaluate the safety of UC-MSC-EV administration to patients with complex perianal fistulas associated with Crohn's disease ($n = 5$). No short- or long-term adverse effects were observed in patients, and neither leucocytosis nor abnormal liver function was reported [61]. In terms of efficacy, at the six-month follow-up, four patients responded to the treatment; however, the trial was ineffective in the presence of fibrotic tissue, indicating that EVs cannot penetrate fibrosis and may therefore be unsuitable for conditions with significant fibrosis. The same research group conducted another clinical trial using EVs derived from placental MSCs in patients with complex perianal fistulas not associated with Crohn's disease ($n = 11$). Ten patients showed positive outcomes, and none experienced any acute or delayed allergic reactions or injection-related complications [62].

Concerning eye pathologies, a pilot study tested the safety and efficacy of UC-MSC-EV intravitreal injections in promoting healing of large and refractory macular holes, which are small gaps that open at the center of the retina, to improve visual outcomes of surgery. Four out of five macular holes closed successfully with MSC-EV therapy, and three patients experienced significant improvement in visual acuity [106].

Various clinical trials with MSC-EVs are still ongoing (Table 7.2). For instance, one such research tests the effects of MSC-EVs on lesions in subjects with epidermolysis bullosa. In particular, the aim of this study is to assess the safety and efficacy of AGLE-102, which is an allogeneic-derived EV product derived from normal donor BM-MSCs (NCT04173650). Multiple organ dysfunction syndrome, instead, is a common debilitating complication after surgical repair for acute type A aortic dissection. In a study, patients undergoing surgical repair of aortic dissection presenting severe multi-organ failure either immediately or after surgical intervention are treated with UC-MSC-EVs, administered intravenously, once a day for 14 times (NCT04356300). In another study, researchers are aiming to assay the administration of MSC-derived exosomes on the improvement of disability of patients with acute ischemic stroke. In this case, MSC-EVs have been engineered in order to be enriched by miR-124 (NCT03384433). Another clinical trial is assessing the effect of intravenous infusion of UCB-MSC-EVs in reducing the inflammatory state in type 1 diabetes mellitus patients. Indeed, this pathology is an autoimmune-mediated disease destructing the islets β-cell of the pancreas and MSC-EVs have been reported to act as anti-inflammatory agents (NCT02138331). Finally, concerning neurodegenerative disorders, researchers are evaluating the safety and efficacy of allogeneic AT-MSC-exosomes in the treatment of mild to moderate dementia due to Alzheimer's disease. In this case, the administration is intranasal, twice a week for 12 weeks (NCT04388982). The clinical trials research data are obtained from trialsearch.who.int, clinicaltrials.gov, and clinicaltrialsregister.eu.

The outbreak of coronavirus disease 2019 (COVID-19) pandemic forced the scientific community to explore new possibilities to contrast this disease, for which there is no effective treatment yet [75]. Seeing the potential of MSC-EVs in other pathologies, many studies have been testing MSC-EVs for the treatment of COVID-19, obtaining very encouraging results also in clinical trials. These reported

Table 7.2 List of ongoing clinical trials on the use of EVs from different MSC sources in various acute and chronic pathologies

References	Status	Trial phase	Disease	EV source	Administration route
NCT04173650	Not yet recruiting	Phases I and II	Dystrophic epidermolysis bullosa	BM-MSCs	Cutaneous
NCT04356300	Not yet recruiting	Not applicable	Multiple organ dysfunction syndrome after surgical repair of acute type A aortic dissection	UC-MSCs	Intravenous
NCT03384433	Unknown status	Phases I and II	Acute ischemic stroke	Engineered MSCs	Intravenous
NCT02138331	Unknown status	Phases II III	Type I diabetes mellitus	UCB-MSCs	Intravenous
NCT04388982	Unknown status	Phases I and II	Alzheimer's disease–associated dementia	AT-MSCs	Nasal

Abbreviations: *BM-MSCs* bone marrow–derived mesenchymal stem/stromal cells, *AT-MSCs* adipose tissue–derived mesenchymal stem/stromal cells, *UC-MSCs* umbilical cord-derived mesenchymal stem/stromal cells, *UCB-MSCs* umbilical cord blood–derived mesenchymal stem/stromal cells. The clinical trials research data is obtained from trialsearch.who.int, clinicaltrials.gov, and clinicaltrialsregister.eu

beneficial effects represent a good example of the therapeutic potential of MSC-EVs. For this reason, in the following sections we will provide an analysis relative to the findings of MSC-EVs in the context of COVID-19.

5 SARS-CoV-2 and COVID-19 Treatment

The severe acute respiratory syndrome β-coronavirus 2 (SARS-CoV-2) was first reported in Wuhan, China, in December 2019, and it is responsible for COVID-19 infectious disease [64, 86].

SARS-CoV-2 belongs to the Coronaviridae family, a group of enveloped viruses with a positive-sense, single-stranded RNA genome [32]. Its structure includes a spike glycoprotein, an M-membrane protein (a type III transmembrane glycoprotein), an N nucleocapsid protein within the phospholipid bilayer, and various non-structural proteins [32]. Compared to other zoonotic coronaviruses such as Middle East respiratory syndrome (MERS-CoV) and swine acute diarrhea syndrome (SADS-CoV), SARS-CoV-2 is more contagious due to its human-to-human transmission via respiratory droplets produced during sneezing, coughing, breathing, and speaking, along with its longer incubation period (2–14 days) [9]. The severity of COVID-19 is largely driven by the host's immune response, with severe cases often leading to acute respiratory distress syndrome (ARDS), characterized by aggressive inflammatory reactions, hyaline membrane formation, and pulmonary fibrosis [1]. The SARS-CoV-2 genome is enclosed by a membrane that contains various viral proteins, most notably the spike (S) protein, which plays a critical role in the virus entrance into target cells [64]. The SARS-CoV-2 infection begins with the virus binding between the spike and angiotensin-converting enzyme 2 (ACE2) receptor on the host cell. ACE2 is expressed in lung alveolar epithelial cells, small intestinal epithelial cells, cardiovascular system, central nervous system, and kidneys [23, 30]. After the spike-ACE2 interaction, the S protein conformation changes, which facilitates virus entry into the endosomal pathway. The serine protease TMPRSS2 is responsible of spike conformation changes, following the S-ACE2 binding, for the internalization of virus by cell and viral membrane fusion [23, 30, 64]. Once the genome is released into the host cytosol, RNA genome is translated into replicase proteins (ORF1a/b). These polyproteins (pp1a and pp1b) are then broken down by a virus-encoded protease into individual nonstructural proteins (nsps), including the RNA-dependent RNA polymerase (RdRp). Replication occurs within double-membrane vesicles (DMVs). The positive-strand RNA genome acts as a template to produce both full-length negative-strand RNA and subgenomic RNA (sgRNA). The sgRNA is translated into structural and accessory proteins (N, S, M, and E), which are inserted into the ER-Golgi intermediate compartment (ERGIC) to create new virions [26]. New SARS-CoV-2 particles are released via exocytosis and leave the host cell to infect new cells [2, 26].

SARS-CoV-2 is a cytopathic virus that can cause injury and death to cells and tissues during its life cycle [63]. Infection and replication in ACE2-positive cells may promote pyroptosis, a type of cell death linked to vascular leakage [1]. Pyroptosis is a caspase-1-driven process of programmed cell death that leads to an inflammatory response [1].

During the virus replication, the host's immune system is activated, inducing the recruitment of inflammatory cells with subsequent production of pro-inflammatory cytokines and chemokines [7]. During virus infection, different pro-inflammatory cytokines and other mediators are significantly increased, like IL-2, IL-10, IL-7, IL-6, IL-1β, IFNγ, INF-α, and monocyte chemoattractant protein-1 (MCP-1), causing risk of multi-organ failure [16, 18, 98].

Conventional treatment for COVID-19 focuses on infection prevention and supportive care, including oxygen supplementation and mechanical ventilation [24]. Currently available therapeutic approaches consist in antivirals and monoclonal antibodies; moreover, various anti-inflammatory and immunomodulatory drugs have been tested to control the cytokine storm and prevent multiple organ failure in severe cases. However, their impact has not been significant [97]. Antiviral or antibody-based treatments are most effective only during the early or replicative state of the disease. In late stage, affected patients may develop pulmonary infiltrates with an age-related final fatality rate [95]. Currently, available therapies mainly focus on target the symptoms caused by the SARS-CoV-2 infection, but further efforts are necessary to advance the development of effective COVID-19 treatments [99, 76]. In the absence of specific antiviral therapies for SARS-CoV-2 and its complications, the primary strategy remains infection prevention through vaccination [100].

6 MSC and MSC-EVs in COVID-19.

There is an increasing number of clinical investigations to find therapeutic solutions to the various forms of the COVID-19 disease. The acute lung injury is the most severe impact of COVID-19 [48]. Host response promotes damage to pulmonary tissue, a reduction in total function, and a reduced lung capacity depending on the lack of control over the immune response [48]. Studies focused on the regeneration of damaged cells as well as on the inhibiting or modulating of inflammatory responses [1, 64]. MSCs and MSC-EVs are being intensively investigated for the treatment of COVID-19 [64]. MSCs may improve damaged lung tissue due to homing in the direction of injuries site of the lung to sustain homeostasis as well as promote regeneration and secretion of paracrine factors, like exosomes, to moderate inflammation and induce tissue regeneration [33, 58, 85]. Recently, a clinical study has been performed using the administration of umbilical cord-derived MSCs in seven COVID-19 patients [72]. Interestingly, the MSC injection significantly improved pulmonary functions and symptoms, as well as changes of inflammatory and immune function level [47]. Another work showed a similar effect in a 65-year-old woman with severe COVID-19. CT images showed remission of the inflammation and good tolerance of allogeneic umbilical cord-MSC transfer [51]. These are findings from a few cohorts of patients and many of therapeutic activities of MSCs are attributed to the paracrine action through the release of EVs [101].

MSC-derived EVs are studied in various aspects of COVID-19, including their use as biomarkers for diagnostic purpose, or, when engineered, as immune-modulating agents, targeted therapeutic cargo, as well as vaccines [1].

MSC-EVs in preclinical models of pulmonary diseases showed the most common effect in an improvement in lung functions [6, 73], which is expressed through the increase of cell survival, the reduction of the inflammatory response and the restoration of the lung architecture, with reduction of fibrosis and increase of alveolarization [6]. A recent work conducted in a mouse model of LPS-induced lung injury showed a reduction of the damage with MSC-EV treatment [73]. Moreover, in another study MSC-EVs have been reported to reduce influenza virus-induced lung damage, including the prevention of apoptosis in lung cells. This protective effect is facilitated through mechanisms such as the production of growth factors, including keratinocyte growth factor, reduction of cytokine and inflammatory cell infiltration, and enhancement of alveolar fluid clearance [37]. The potential of MSC-EVs in repairing and regenerating tissue is highly relevant considering the lesions reported after SARS-CoV-2 infection. MSC-EVs have been shown to stimulate the repopulation of type II alveolar cells, which act as progenitor cells for lung epithelium regeneration [42].

In addition, in experimental bronchopulmonary dysplasia model, the MSC-EVs restore lung function through macrophage immunomodulation [92]. The benefits of using MSC-EVs were showed also in allergic airway inflammation models [34]. Moreover, MSC-EVs were explored in hemorrhagic shock and trauma, as well as in severe pneumonia, lung ischemic reperfusion, and lung fibrosis [59, 70, 77].

In COVID-19 subjects, the mortality result increased in patients with diabetes mellitus or renal disease [36]. Data indicate that high glucose levels and uremic toxins exacerbate inflammation [36]. Notably, the incorporation of Wharton's jelly MSC-derived EVs into SARS-CoV peptide-induced Calu-3 cells led to a significant reduction in nuclear NF-κB p65 levels and downregulated the cytokine storm, even under conditions of elevated glucose and uremic toxins [36].

The molecular mechanisms and the key factors involved in MSC-EVs therapeutic effects are, for the most, still undiscovered [91]. Culture conditions can strongly affect the EV biological effects. In fact, EV composition may vary depending on several factors, such as donor's condition, culture conditions, or the microenvironment. It is possible to act on the microenvironment to "educate" the cells to produce the best EVs in relation to the therapeutic need [91]. In the context of pulmonary disease, the ability of conditioned media on EV from MSCs was demonstrated, for example, by Keyhanmanesh et al. [34], and Li et al. [49].

7 EVs as Biomarkers in COVID-19

EVs from MSCs were known to contain heterogeneous cargo, including functional microRNAs (miRNAs), small noncoding sequences implicated in post-transcriptional gene repression or degradation, and can attach various targets [67]. Schultz et al. reported the property of MSC-EVs miRNAs to regulate the cytokines levels, cell death and coagulation, factors involved in COVID-19 [67].

The results described those 258 miRNAs that were concerned with the cytokines and chemokine increase, 266 miRNAs with cell death and 148 miRNAs with coagulation cascades. On the other side, the study recognized some miRNAs able to reduce inflammatory agents and inhibit cell death genes and key factors of coagulation cascade [67]. However, few studies investigated the correlation of MSC-EV-cargo molecules with several components of SARS-CoV-2. Recently, it was demonstrated that EVs from plasma patients with COVID-19 are enriched in cardiovascular, inflammatory, and procoagulant proteins according to the severity of disease [40]. Balbi et al. [5] showed high activity of CD142, procoagulant tissue factor, on EV surface in COVID-19 patients, correlated with TNF-α serum levels [5]. Moreover, it was found the increasing of COPB2 protein, a subunit of the Golgi coatomer complex, associated to EVs in patients remained mild than developed severe/critical COVID-19 [21]. Another study investigated the changes in EV composition in patients at different temporal stages of COVID-19 (the pre-symptomatic, hyperinflammatory, resolution, and convalescent phases) [44]. The results revealed a dysregulated raft lipid metabolism that underlies changes in EV lipid membrane anisotropy that alter the exosomal localization of presenilin-1 (PS-1) in the hyperinflammatory phase [44]. In addition, they showed in vitro that EVs, from different temporal phases, trigger distinct metabolic and transcriptional responses in alveolar epithelial and in liver hepatocytes cells [44]. In comparison to the hyperinflammatory phase, EVs from the resolution phase induce opposite effects on eukaryotic translation and Notch signaling [44]. Also, the miRNAs analysis of EVs from patients with COVID-19 pneumonia and patients with severe COVID-19 demonstrated that they are connected to the regulation of signaling pathway involved in the progression of disease [55]. Wang et al. identified miR7-5p, miR24-3p, miR145-5p, and miR223-3p in serum EVs that can inhibit spike protein expression and SARS-CoV-2 replication. These miRNAs were remarkably decreased in the elderly and people with diabetes [88]. These findings highlight the potential of EV-cargo or EV-associated molecules as biomarkers for patient stratification and for making early clinical decisions about strategies for COVID-19 therapy.

8 Engineered EVs and COVID-19

There is a growing interest in the new potential application of EV as a disease treatment strategy [1, 64]. In particular, the engineering of EVs could be attractive for antiviral purposes. Recent studies demonstrated that ACE2-engineered EVs limit the SARS-CoV-2 infection [17, 104, 105]. Interestingly, exosomes enriched in ACE2 protein were developed to induce a competitive inhibition-mediated binding of the exosomes to the SARS-CoV-2 spike protein [17, 104, 105]. Alternatively, Scott et al. propose a molecular therapy to neutralize SARS-CoV-2 virus using EVs containing a fusion protein, in particular, a CD63 tetraspanin linked to an anti-SARS-CoV-2 nanobody. These EVs, derived from HEK293 cells and enriched with an anti-SARS-CoV-2 antibody, bind to the receptor-binding domain (RBD) of the viral spike protein, effectively neutralizing the virus [68].

In 2022, one research showed that EVs expressing spike protein on their surface were a safe method for the study of COVID-19. In particular, the reduction of virus-like particles entrance in the host cells was demonstrated in presence of new possible agents (colchicine or anti-ACE2) against SARS-CoV-2 infection [86]. Another study demonstrated the potential of using EVs modified with the receptor-binding domain (RBD) of the SARS-CoV-2-spike protein, which binds to the ACE2 receptor, as a targeted delivery system for potential antiviral agents in vivo [20, 87].

9 EVs Vaccines for COVID-19

The use of EVs as immunogenicity factors in the treatment of SARS-CoV infection has been studied [65, 71, 97].

Kuate et al. analyzed exosome-based vaccines containing spike protein by replacing the protein transmembrane and cytoplasmic domains with those of glycoprotein G of vesicular stomatitis virus (VSV-G), and its immunogenicity and efficacy were tested in mice [41].

In addition, it has been proved that both exosomes and the vaccine, with adenoviral vector expressing the S protein, could induce neutralizing antibodies [14, 97].

Moreover, the EV treatment has been shown to be more efficient than soluble protein subunit vaccines, since EVs can express several copies of the same viral protein and promote the cross-linking between EVs and the B cell receptor [25].

In contrast, Troyer Z et al. demonstrated that EVs containing SARS-CoV-2 spike interact with the humoral immune system and decrease serum neutralizing antibodies of convalescent patients [82].

10 Clinical Trials on MSC-EVs in the Treatment of COVID-19

Due to the rapid spread of COVID-19 and the absence of an efficient therapy, several clinical studies were approved by ethical committees [39]. Some of these focused on EVs from MSCs as a new therapeutic strategy for COVID-19 (Table 7.3). Recently, MSC-derived EVs were administered, by single intravenous injection, to 24 patients with moderate to severe ARDS. Adverse effects were not observed and the survival rate was 83% [69]. Also, the amniotic fluid-derived EVs were effective with important respiratory improvements in COVID-19 patients [56, 57].

Among the completed clinical trials, two studies investigated the inhalation of EVs-based formulation. EVs were administered at the concentrations of 0.5-2$\times 10^{10}$ particles/mL, twice a day for 10 days (NCT04491240); in the other study, instead, at the concentration of 2×10^{8} particles/mL, once a day for 5 days (NCT04276987). Other clinical trials focused on bone marrow MSC–derived EVs (BM-MSC-EVs) used in patients with ARDS (NCT04493242 and NCT04798716). A study evaluated the safety and efficacy of intravenous administration of BM-MSC-EVs versus placebo as a treatment for COVID-19 patients with moderate-to-severe ARDS. The EV administration resulted in safety in patients with severe or critical COVID-19 respiratory failure and the mortality was reduced compared to the placebo [52]. Another trial focused on a scaling dose of MSC-EV administration by intravenous injection (phase I: 2×10^{9}, 4×10^{9}, 8×10^{9} particles/mL; phase II: 4×10^{9}, 8×10^{9} particles/mL) for a period of 5 days (NCT04798716). In Iran, a double-blinded randomized controlled clinical trial was conducted on patients with COVID-19-associated ARDS, using human placental MSC-EVs. About 1.5-2 $\times 10^{9}$ EVs/kg were administered intravenously for two consecutive days (IRCT20130812014333N164) [102]. The authors observed a significant decrease in the mortality rate in patients treated with MSC-EVs compared to the control group [102]. Moreover, various clinical trials on MSC-EVs in COVID-19 are still ongoing or recruiting, as reported in Table 7.3. Also, engineered EVs were tested in COVID-19 clinical trials. In particular, one study investigated the inhalation of EVs that overexpress CD24 from human embryonic kidney cell, one daily for 5 days, in phase II (NCT04969172 and NCT04747574). Another trial used exosomes derived from T cells that were activated in vitro with COVID-19 viral peptide fragments (2×10^{8}/mL of EVs five times daily for 5 days; phase I; NCT04389385). Another clinical trial for COVID-19

Table 7.3 List of clinical trials on the use of EVs from MSC sources in COVID-19

References	Status	Trial phase	COVID-19	EV source	Administration route
NCT046657458	Completed	Phase II	ARDS	BM-MSCs	Intravenous
NCT04491240	Completed	Phases I and II	Pneumonia	MSCs	Inhalation
NCT04276987	Completed	Phase I	Pneumonia	AT-MSCs	Inhalation
NCT04493242	Completed	Phase II	ARDS	BM-MSCs	Intravenous
IRCT20130812014333N164	Completed	Phases II and III	ARDS	P-MSCs	Intravenous
ISRCTN33578935	Ongoing	Phase II	ARDS	P-MSCs	Intravenous
IRCT20190101042197N2	Recruiting	Phases I and II	ARDS	MSCs	Intravenous
IRCT20201202049568N3	Recruiting	Phases I and II	ARDS	UC-MSCs	Intravenous
IRCT20200413047063N2	Recruiting	Phases I and II	ARDS	P-MSCs	Intravenous
NCT04798716	Not yet recruiting	Phases I and II	Pneumonia	BM-MSCs	Intravenous
NCT05354141	Recruiting	Phase III	ARDS	BM-MSCs	Intravenous
NCT05216562	Recruiting	Phases II and III	SARS-CoV-2 infection	MSCs	Intravenous
NCT05116761	Not yet recruiting	Phases I and II	Post-acute and chronic post-COVID-19	BM-MSCs	Intravenous
NCT05808400	Recruiting	Phase I	Chronic cough	UC-MSCs	Inhalation
NCT05125562	Withdrawn	Phase II	Mild-to-moderate COVID-19	BM-MSCs	Intravenous
NCT04657458	Ongoing	Phase II	ARDS	BM-MSCs	Intravenous
NCT05787288	Recruiting	Phase I	Pneumonia	UC-MSCs	Inhalation
NCT05127122	Withdrawn	Phases I and II	ARDS	BM-MSCs	Intravenous

Abbreviations: *BM-MSCs* bone marrow–derived mesenchymal stem/stromal cells, *AT-MSCs* adipose tissue–derived mesenchymal stem/stromal cells, *UC-MSCs* umbilical cord–derived mesenchymal stem/stromal cells, *CB-MSCs* cord blood–derived mesenchymal stem/stromal cells, *P-MSCs* placenta-derived mesenchymal stem/stromal cells. The clinical trials research data is obtained from trialsearch.who.int, clinicaltrials.gov, and clinicaltrialsregister.eu used Zofin, an acellular product derived from human amniotic fluid, that contain growth factors, cytokines, chemokines, and EVs (2–5 × 10^{11} particles/mL, intravenous administration on day 0, day 4, and day 8; phases I and II; NCT04384445). However, the exact composition and EV content and identity are unclear.

The clinical trials research data are obtained from trialsearch.who.int, clinicaltrials.gov, and clinicaltrialsregister.eu.

11 Conclusions

The administration of MSC-EVs has demonstrated positive outcomes in both preclinical and clinical studies across a wide range of diseases. Herein, we described some of these studies, with a particular focus on COVID-19 treatment, where MSC-EVs appeared to aid in the recovery of lung function and in reducing the lung fibrosis, mainly due to their immunomodulatory effects, thereby promoting lung regeneration.

The International Society for Cellular and Gene Therapies (ISCT) and the International Society for Extracellular Vesicles (ISEV) recognized the potential of MSC-EVs as treatment for COVID-19, encouraging further research and trials in this field [10]. At the same time, ISCT and ISEV recommend the standardization of protocols, including massive production, isolation, characterization, functional evaluation, and batch-to-batch consistency to take advantage of the regenerative potential of MSC-EVs [10]. Further studies are needed to better investigate the MSC-EV mechanism of action, as well as their immunosuppressive effects in the long term.

Conflict of Interest The authors declare no conflict of interest.

References

1. Akbari A, Rezaie J. Potential therapeutic application of mesenchymal stem cell-derived exosomes in SARS-CoV-2 pneumonia. Stem Cell Res Ther. 2020;11(1):1–10. https://doi.org/10.1186/s13287-020-01866-6.
2. Alanagreh L, Alzoughool F, Atoum M. The human coronavirus disease covid-19: its origin, characteristics, and insights into potential drugs and its mechanisms. Pathogens. 2020;9(5) https://doi.org/10.3390/pathogens9050331.
3. Andrzejewska A, Lukomska B, Janowski M. Concise review: mesenchymal stem cells: from roots to boost. Stem Cells (Dayton, Ohio). 2019;37(7):855–64. https://doi.org/10.1002/stem.3016.
4. Arslan F, et al. Mesenchymal stem cell-derived exosomes increase ATP levels, decrease oxidative stress and activate PI3K/Akt pathway to enhance myocardial viability and prevent adverse remodeling after myocardial ischemia/reperfusion injury. Stem Cell Res. 2013;10(3):301–12. https://doi.org/10.1016/j.scr.2013.01.002.

5. Balbi C, et al. Circulating extracellular vesicles are endowed with enhanced procoagulant activity in SARS-CoV-2 infection. EBioMedicine. 2021;67:103369. https://doi.org/10.1016/j.ebiom.2021.103369.
6. Bari E, et al. Mesenchymal stem/stromal cell secretome for lung regeneration: the long way through "pharmaceuticalization" for the best formulation. J Control Release. 2019;309(April):11–24. https://doi.org/10.1016/j.jconrel.2019.07.022.
7. Batah SS, Fabro AT. Pulmonary pathology of ARDS in COVID-19: a pathological review for clinicians. Respir Med. 2021;176:106239. https://doi.org/10.1016/j.rmed.2020.106239.
8. Bian S, et al. Extracellular vesicles derived from human bone marrow mesenchymal stem cells promote angiogenesis in a rat myocardial infarction model. J Mol Med (Berl). 2014;92(4):387–97. https://doi.org/10.1007/s00109-013-1110-5.
9. Bimonte S, et al. Potential antiviral drugs for SARS-Cov-2 treatment: preclinical findings and ongoing clinical research. In Vivo. 2020;34:1597–602. https://doi.org/10.21873/invivo.11949.
10. Börger V, et al. International Society for Extracellular Vesicles and International Society for cell and gene therapy statement on extracellular vesicles from mesenchymal stromal cells and other cells: considerations for potential therapeutic agents to suppress coronavirus disease-19. Cytotherapy. 2020;22(9):482–5. https://doi.org/10.1016/j.jcyt.2020.05.002.
11. Bruno S, et al. Mesenchymal stem cell-derived microvesicles protect against acute tubular injury. J Am Soc Nephrol. 2009;20(5):1053–67. https://doi.org/10.1681/ASN.2008070798.
12. Bruno S, et al. Microvesicles derived from mesenchymal stem cells enhance survival in a lethal model of acute kidney injury. PLoS One. 2012;7(3):e33115. https://doi.org/10.1371/journal.pone.0033115.
13. Bruno S, et al. Renal regenerative potential of different extracellular vesicle populations derived from bone marrow mesenchymal stromal cells. Tissue Eng Part A. 2017;23(21–22):1262–73. https://doi.org/10.1089/ten.TEA.2017.0069.
14. Cacciottolo M, et al. Nanograms of SARS-CoV-2 spike protein delivered by exosomes induce potent neutralization of both delta and omicron variants. PLoS One. 2023;18(8):e0290046. https://doi.org/10.1371/journal.pone.0290046.
15. Caplan AI. Mesenchymal stem cells: time to change the name! Stem Cells Transl Med. 2017;6(6):1445–51. https://doi.org/10.1002/sctm.17-0051.
16. Chen G, et al. Clinical and immunological features of severe and moderate coronavirus disease 2019. J Clin Invest. 2020;130(5):2620–9. https://doi.org/10.1172/JCI137244.
17. Cocozza F, et al. Extracellular vesicles containing ACE2 efficiently prevent infection by SARS-CoV-2 spike protein-containing virus. J Extracell Vesicles. 2020;10(2):e12050. https://doi.org/10.1002/jev2.12050.
18. Diao B, et al. Reduction and functional exhaustion of T cells in patients with coronavirus disease 2019 (COVID-19). Front Immunol. 2020;11(May):1–7. https://doi.org/10.3389/fimmu.2020.00827.
19. Dominici M, et al. Minimal criteria for defining multipotent mesenchymal stromal cells. The International Society for Cellular Therapy position statement. Cytotherapy. 2006;8(4):315–7. https://doi.org/10.1080/14653240600855905.
20. Fu Y, Xiong S. Tagged extracellular vesicles with the RBD of the viral spike protein for delivery of antiviral agents against SARS-COV-2 infection. J Control Release. 2021;335:584–95. https://doi.org/10.1016/j.jconrel.2021.05.049.
21. Fujita Y, et al. Early prediction of COVID-19 severity using extracellular vesicle COPB2. J Extracell Vesicles. 2021;10(8):e12092. https://doi.org/10.1002/jev2.12092.
22. Gatti S, et al. Microvesicles derived from human adult mesenchymal stem cells protect against ischaemia-reperfusion-induced acute and chronic kidney injury. Nephrol Dial Transplant. 2011;26(5):1474–83. https://doi.org/10.1093/ndt/gfr015.
23. Gheblawi M, et al. Angiotensin-converting enzyme 2: SARS-CoV-2 receptor and regulator of the renin-angiotensin system: celebrating the 20th anniversary of the discovery of ACE2. Circ Res. 2020;126(10):1456–74. https://doi.org/10.1161/CIRCRESAHA.120.317015.

24. Grasselli G, et al. Baseline characteristics and outcomes of 1591 patients infected with SARS-CoV-2 admitted to ICUs of the Lombardy region, Italy. JAMA. 2020;323(16):1574–81. https://doi.org/10.1001/jama.2020.5394.
25. György B, et al. Therapeutic applications of extracellular vesicles: clinical promise and open questions. Annu Rev Pharmacol Toxicol. 2015;55:439–64. https://doi.org/10.1146/annurev-pharmtox-010814-124630.
26. Harrison AG, Lin T, Wang P. Mechanisms of SARS-CoV-2 transmission and pathogenesis. Trends Immunol. 2020;41(12):1100–15. https://doi.org/10.1016/j.it.2020.10.004.
27. Haynesworth SE, Baber MA, Caplan AI. Cytokine expression by human marrow-derived mesenchymal progenitor cells in vitro: effects of dexamethasone and IL-1 alpha. J Cell Physiol. 1996;166(3):585–92.
28. He J, et al. Bone marrow stem cells-derived microvesicles protect against renal injury in the mouse remnant kidney model. Nephrology. 2012;17(5):493–500. https://doi.org/10.1111/j.1440-1797.2012.01589.x.
29. Horwitz EM, et al. Clarification of the nomenclature for MSC: the International Society for Cellular Therapy position statement. Cytotherapy. 2005;7(5):393–5. Available at:. https://doi.org/10.1080/14653240500319234.
30. Hu B, et al. Characteristics of SARS-CoV-2 and COVID-19. Nat Rev Microbiol. 2021;19(3):141–54. https://doi.org/10.1038/s41579-020-00459-7.
31. Kanazawa H, et al. Bone marrow-derived mesenchymal stem cells ameliorate hepatic ischemia reperfusion injury in a rat model. PLoS One. 2011;6(4):e19195. https://doi.org/10.1371/journal.pone.0019195.
32. Kaul D. An overview of coronaviruses including the SARS-2 coronavirus – molecular biology, epidemiology and clinical implications. Curr Med Res Pract. 2020;10(3):135. https://doi.org/10.1016/j.cmrp.2020.04.001.
33. Keshtkar S, et al. Mesenchymal stem cell-derived extracellular vesicles: novel frontiers in regenerative medicine. Stem Cell Res Ther. 2018;9(1):63. https://doi.org/10.1186/s13287-018-0791-7.
34. Keyhanmanesh R, Rahbarghazi R, Ahmadi M. Systemic transplantation of mesenchymal stem cells modulates endothelial cell adhesion molecules induced by ovalbumin in rat model of asthma. Inflammation. 2018;41(6):2236–45. https://doi.org/10.1007/s10753-018-0866-8.
35. Khademizadeh M, et al. Differentiation of adult human mesenchymal stem cells into dopaminergic neurons. Res in Pharm Sci. 2019;14(3):209–15. https://doi.org/10.4103/1735-5362.258487.
36. Khanh VC, et al. Wharton's jelly mesenchymal stem cell-derived extracellular vesicles reduce SARS-CoV2-induced inflammatory cytokines under high glucose and uremic toxin conditions. Stem Cells Dev. 2021;30(15):758–72. https://doi.org/10.1089/scd.2021.0065.
37. Khatri M, Richardson LA, Meulia T. Mesenchymal stem cell-derived extracellular vesicles attenuate influenza virus-induced acute lung injury in a pig model. Stem Cell Res Ther. 2018;9(1):1–13. https://doi.org/10.1186/s13287-018-0774-8.
38. Kordelas L, et al. MSC-derived exosomes: a novel tool to treat therapy-refractory graft-versus-host disease. Leukemia. 2014;28(4):970–3. https://doi.org/10.1038/leu.2014.41.
39. Krishnan A, et al. Mesenchymal stem cell-derived extracellular vesicles in the management of COVID19-associated lung injury: a review on publications, clinical trials and patent landscape. Tissue Eng Regen Med. 2022;19:659. https://doi.org/10.1007/s13770-022-00441-9.
40. Krishnamachary B, Cook C, Kumar A, Spikes L, Chalise P, Dhillon NK. Extracellular vesicle-mediated endothelial apoptosis and EV-associated proteins correlate with COVID-19 disease severity. J Extracell Vesicles. 2021;10(9):e12117. https://doi.org/10.1002/jev2.12117. Epub 2021. PMID: 34262673; PMCID: PMC8254805.
41. Kuate S, et al. Exosomal vaccines containing the S protein of the SARS coronavirus induce high levels of neutralizing antibodies. Virology. 2007;362(1):26–37. https://doi.org/10.1016/j.virol.2006.12.011.

42. Laffey JG, Matthay MA. Cell-based therapy for acute respiratory distress syndrome: biology and potential therapeutic value. Am J Respir Crit Care Med. 2017;196(3):266–73. https://doi.org/10.1164/rccm.201701-0107CP.
43. Lai RC, et al. Exosome secreted by MSC reduces myocardial ischemia/reperfusion injury. Stem Cell Res. 2010;4(3):214–22. https://doi.org/10.1016/j.scr.2009.12.003.
44. Lam SM, et al. A multi-omics investigation of the composition and function of extracellular vesicles along the temporal trajectory of COVID-19. Nat Metab. 2021;3(7):909–22. https://doi.org/10.1038/s42255-021-00425-4.
45. Lee C, et al. Exosomes mediate the cytoprotective action of mesenchymal stromal cells on hypoxia-induced pulmonary hypertension. Circulation. 2012;126(22):2601–11. https://doi.org/10.1161/CIRCULATIONAHA.112.114173.
46. Lee JW, et al. Allogeneic human mesenchymal stem cells for treatment of E. coli endotoxin-induced acute lung injury in the ex vivo perfused human lung. Proc Natl Acad Sci USA. 2009;106(38):16357–62. https://doi.org/10.1073/pnas.0907996106.
47. Leng Z, et al. Transplantation of ACE2- mesenchymal stem cells improves the outcome of patients with covid-19 pneumonia. Aging Dis. 2020;11(2):216–28. https://doi.org/10.14336/AD.2020.0228.
48. Li G, et al. Coronavirus infections and immune responses. J Med Virol. 2020;92(4):424–32. https://doi.org/10.1002/jmv.25685.
49. Li L, Jin S, Zhang Y. Ischemic preconditioning potentiates the protective effect of mesenchymal stem cells on endotoxin-induced acute lung injury in mice through secretion of exosome. Int J Clin Exp Med. 2015;8(3):3825–32.
50. Li T, et al. Exosomes derived from human umbilical cord mesenchymal stem cells alleviate liver fibrosis. Stem Cells Dev. 2013;22(6):845–54. https://doi.org/10.1089/scd.2012.0395.
51. Liang B, et al. Clinical remission of a critically ill COVID-19 patient treated by human umbilical cord mesenchymal stem cells: a case report. Medicine. 2020;99(31):e21429. https://doi.org/10.1097/MD.0000000000021429.
52. Lightner AL, et al. Bone marrow mesenchymal stem cell-derived extracellular vesicle infusion for the treatment of respiratory failure from COVID-19: a randomized, placebo-controlled dosing clinical trial. Chest. 2023;164(6):1444–53. https://doi.org/10.1016/j.chest.2023.06.024.
53. Maqsood M, et al. Adult mesenchymal stem cells and their exosomes: sources, characteristics, and application in regenerative medicine. Life Sci. 2020;256:118002. https://doi.org/10.1016/j.lfs.2020.118002.
54. Mazini L, et al. Regenerative capacity of adipose derived stem cells (ADSCs), comparison with mesenchymal stem cells (MSCs). Int J Mol Sci. 2019;20(10):E2523. https://doi.org/10.3390/ijms20102523.
55. Meidert AS, et al. Extracellular vesicle associated miRNAs regulate signaling pathways involved in COVID-19 pneumonia and the progression to severe acute respiratory Corona Virus-2 syndrome. Front Immunol. 2021;12(December):1–16. https://doi.org/10.3389/fimmu.2021.784028.
56. Mitrani MI, Bellio MA, Meglin A, et al. Treatment of a COVID-19 long hauler with an amniotic fluid-derived extracellular vesicle biologic. Respir Med Case Rep. 2021a;34:101502. https://doi.org/10.1016/j.rmcr.2021.101502.
57. Mitrani MI, Bellio MA, Sagel A, et al. Case report: Administration of Amniotic Fluid-Derived Nanoparticles in three severely ill COVID-19 patients. Front Med. 2021b;8(March):1–8. https://doi.org/10.3389/fmed.2021.583842.
58. Mohammadipoor A, et al. Therapeutic potential of products derived from mesenchymal stem/stromal cells in pulmonary disease. Respir Res. 2018;19(1):1–14. https://doi.org/10.1186/s12931-018-0921-x.
59. Monsel A, et al. Therapeutic effects of human mesenchymal stem cell-derived microvesicles in severe pneumonia in mice. Am J Respir Crit Care Med. 2015;192(3):324–36. https://doi.org/10.1164/rccm.201410-1765OC.

60. Nassar W, et al. Umbilical cord mesenchymal stem cells derived extracellular vesicles can safely ameliorate the progression of chronic kidney diseases. Biomater Res. 2016;20:21. https://doi.org/10.1186/s40824-016-0068-0.
61. Nazari H, et al. Evaluating the safety and efficacy of mesenchymal stem cell-derived exosomes for treatment of refractory perianal fistula in IBD patients: clinical trial phase I. Gastroenterol Rep. 2022;10 https://doi.org/10.1093/gastro/goac075.
62. Pak H, et al. Safety and efficacy of injection of human placenta mesenchymal stem cells derived exosomes for treatment of complex perianal fistula in non-Crohn's cases: clinical trial phase I. J Gastroenterol Hepatol. 2023;38(4):539–47. https://doi.org/10.1111/jgh.16110.
63. Park WB, et al. Virus isolation from the first patient with SARS-CoV-2 in Korea. J Korean Med Sci. 2020;35(7):e84. https://doi.org/10.3346/jkms.2020.35.e84.
64. Pocsfalvi G, et al. COVID-19 and extracellular vesicles: an intriguing interplay. Kidney Blood Press Res. 2020;45:661. https://doi.org/10.1159/000511402.
65. Sabanovic B, Piva F, Cecati M, Giulietti M. Promising extracellular vesicle-based vaccines against viruses, including SARS-CoV-2. Biology. 2021;10(2):94. https://doi.org/10.3390/biology10020094.
66. Samsonraj RM, et al. Concise review: multifaceted characterization of human mesenchymal stem cells for use in regenerative medicine. Stem Cells Transl Med. 2017;6(12):2173–85. https://doi.org/10.1002/sctm.17-0129.
67. Schultz IC, Bertoni APS, Wink MR. Mesenchymal stem cell-derived extracellular vesicles carrying miRNA as a potential multi target therapy to COVID-19: an in Silico analysis. Stem Cell Rev Rep. 2021;17(2):341–56. https://doi.org/10.1007/s12015-021-10122-0.
68. Scott TA, Supramaniam A, Idris A, Cardoso AA, Shrivastava S, Kelly G, Grepo NA, Soemardy C, Ray RM, McMillan NAJ, Morris KV. Engineered extracellular vesicles directed to the spike protein inhibit SARS-CoV-2. Mol Ther Methods Clin Dev. 2022;24:355–366. https://doi.org/10.1016/j.omtm.2022.01.015. Epub 2022. PMID: 35127966; PMCID: PMC8806709.
69. Sengupta V, et al. Exosomes derived from bone marrow mesenchymal stem cells as treatment for severe COVID-19. Stem Cells Dev. 2020;29(12):747–54. https://doi.org/10.1089/scd.2020.0080.
70. Shen Q, et al. Paracrine factors from mesenchymal stem cells attenuate epithelial injury and lung fibrosis. Mol Med Rep. 2015;11(4):2831–7. https://doi.org/10.3892/mmr.2014.3092.
71. Shim W, Lee A, Lee JH. The role of extracellular vesicles in pandemic viral infections. J Microbiol (Seoul, Korea). 2024;62(6):419–27. https://doi.org/10.1007/s12275-024-00144-x.
72. Shu L, et al. Treatment of severe COVID-19 with human umbilical cord mesenchymal stem cells. Stem Cell Res Ther. 2020;11(1):1–11. https://doi.org/10.1186/s13287-020-01875-5.
73. Silva JD, et al. Mesenchymal stromal cells are more effective than their extracellular vesicles at reducing lung injury regardless of acute respiratory distress syndrome etiology. Stem Cells Int. 2019;2019:1. https://doi.org/10.1155/2019/8262849.
74. Skovronova R, et al. Surface marker expression in small and medium/large mesenchymal stromal cell-derived extracellular vesicles in naive or apoptotic condition using orthogonal techniques. Cells. 2021;10(11):2948. https://doi.org/10.3390/cells10112948.
75. Sohrabi C, Mathew G, Franchi T, Kerwan A, Griffin M, Soleil C Del Mundo J, Ali SA, Agha M, Agha R. Impact of the coronavirus (COVID-19) pandemic on scientific research and implications for clinical academic training - A review. Int J Surg. 2021;86:57–63. https://doi.org/10.1016/j.ijsu.2020.12.008. Epub 2021. PMID: 33444873; PMCID: PMC7833269.
76. Sreepadmanabh M, Sahu AK, Chande A. COVID-19: advances in diagnostic tools, treatment strategies, and vaccine development. J Biosci. 2020;45(1):148. https://doi.org/10.1007/s12038-020-00114-6.
77. Stone ML, et al. Mesenchymal stromal cell-derived extracellular vesicles attenuate lung ischemia-reperfusion injury and enhance reconditioning of donor lungs after circulatory death. Respir Res. 2017;18(1):1–12. https://doi.org/10.1186/s12931-017-0704-9.

78. Tan CY, et al. Mesenchymal stem cell-derived exosomes promote hepatic regeneration in drug-induced liver injury models. Stem Cell Res Ther. 2014;5(3):76. https://doi.org/10.1186/scrt465.
79. Tatullo M, et al. Exosomes from human periapical cyst-MSCs: Theranostic application in Parkinson's disease. Int J Med Sci. 2020;17(5):657–63. https://doi.org/10.7150/ijms.41515.
80. Teixeira FG, Carvalho MM, Sousa N, Salgado AJ. Mesenchymal stem cells secretome: a new paradigm for central nervous system regeneration? Cell Mol Life Sci. 2013;70(20):3871–82. https://doi.org/10.1007/s00018-013-1290-8.
81. Théry C, et al. Minimal information for studies of extracellular vesicles 2018 (MISEV2018): a position statement of the International Society for Extracellular Vesicles and update of the MISEV2014 guidelines. J Extracell Vesicles. 2018;7(1):1535750. https://doi.org/10.1080/20013078.2018.1535750.
82. Troyer Z, et al. Extracellular vesicles carry SARS-CoV-2 spike protein and serve as decoys for neutralizing antibodies. J Extracell Vesicles. 2021;10(8):e12112. https://doi.org/10.1002/jev2.12112.
83. Tsiapalis D, O'Driscoll L. Mesenchymal stem cell derived extracellular vesicles for tissue engineering and regenerative medicine applications. Cells. 2020a;9(4):E991. https://doi.org/10.3390/cells9040991.
84. Tsiapalis D, O'Driscoll L. Mesenchymal stem cell derived extracellular vesicles for tissue engineering and regenerative medicine applications. Cells. 2020b;9(4) https://doi.org/10.3390/cells9040991.
85. Ullah M, Liu DD, Thakor AS. Mesenchymal stromal cell homing: mechanisms and strategies for improvement. iScience. 2019;15:421–38. https://doi.org/10.1016/j.isci.2019.05.004.
86. Verta R, et al. Generation of spike-extracellular vesicles (S-EVs) as a tool to mimic SARS-CoV-2 interaction with host cells. Cells. 2022;11(1):1–16. https://doi.org/10.3390/cells11010146.
87. Wan Y, et al. Receptor recognition by the novel coronavirus from Wuhan: an analysis based on decade-long structural studies of SARS coronavirus. J Virol. 2020;94(7):1–9. https://doi.org/10.1128/jvi.00127-20.
88. Wang Y, et al. Decreased inhibition of exosomal miRNAs on SARS-CoV-2 replication underlies poor outcomes in elderly people and diabetic patients. Signal Transduct Target Ther. 2021;6(1):300. https://doi.org/10.1038/s41392-021-00716-y.
89. Warnecke A, et al. First-in-human intracochlear application of human stromal cell-derived extracellular vesicles. J Extracell Vesicles. 2021;10(8):e12094. https://doi.org/10.1002/jev2.12094.
90. Welsh JA, et al. Minimal information for studies of extracellular vesicles (MISEV2023): from basic to advanced approaches. J Extracell Vesicles. 2024;13(2):e12404. https://doi.org/10.1002/jev2.12404.
91. Wiklander OPB, et al. Advances in therapeutic applications of extracellular vesicles. Sci Transl Med. 2019;11(492):eaav8521. https://doi.org/10.1126/scitranslmed.aav8521.
92. Willis GR, et al. Mesenchymal stromal cell exosomes ameliorate experimental bronchopulmonary dysplasia and restore lung function through macrophage immunomodulation. Am J Respir Crit Care Med. 2018;197(1):104–16. https://doi.org/10.1164/rccm.201705-0925OC.
93. Willms E, et al. Cells release subpopulations of exosomes with distinct molecular and biological properties. Sci Rep. 2016;6:22519. https://doi.org/10.1038/srep22519.
94. Witwer KW, et al. Defining mesenchymal stromal cell (MSC)-derived small extracellular vesicles for therapeutic applications. J Extracell Vesicles. 2019;8(1):1609206. https://doi.org/10.1080/20013078.2019.1609206.
95. Wu Z, McGoogan JM. Characteristics of and important lessons from the coronavirus disease 2019 (COVID-19) outbreak in China: summary of a report of 72314 cases from the Chinese Center for Disease Control and Prevention. JAMA. 2020;323(13):1239–42. https://doi.org/10.1001/jama.2020.2648.

96. Xu R, et al. Highly-purified exosomes and shed microvesicles isolated from the human colon cancer cell line LIM1863 by sequential centrifugal ultrafiltration are biochemically and functionally distinct. Methods (San Diego, Calif). 2015;87:11–25. https://doi.org/10.1016/j.ymeth.2015.04.008.
97. Yan YY, et al. The potential role of extracellular vesicles in COVID-19 treatment: opportunity and challenge. Front Mol Biosci. 2021;8(July):1–15. https://doi.org/10.3389/fmolb.2021.699929.
98. Yang Y, et al. Plasma IP-10 and MCP-3 levels are highly associated with disease severity and predict the progression of COVID-19. J Allergy Clin Immunol. 2020;146(1):119–127.e4. https://doi.org/10.1016/j.jaci.2020.04.027.
99. Yuan W, Liu S, Lu L, Feng J, He X. Clinical interventions for severe and critical COVID-19: what are the options. Am J Transl Res. 2020;12(5):2110–2117. PMID: 32509205; PMCID: PMC7270004.
100. Yuan Y, et al. The development of COVID-19 treatment. Front Immunol. 2023;14:1125246. https://doi.org/10.3389/fimmu.2023.1125246.
101. Yu B, et al. Exosomes derived from MSCs ameliorate retinal laser injury partially by inhibition of MCP-1. Sci Rep. 2016;6(March):1–12. https://doi.org/10.1038/srep34562.
102. Zamanian MH, et al. Human placental mesenchymal stromal cell-derived small extracellular vesicles as a treatment for severe COVID-19: a double-blind randomized controlled clinical trial. J Extracell Vesicles. 2024;13(7):e12492. https://doi.org/10.1002/jev2.12492.
103. Zhang B, et al. HucMSC-exosome mediated-Wnt4 signaling is required for cutaneous wound healing. Stem Cells (Dayton, Ohio). 2015;33(7):2158–68. https://doi.org/10.1002/stem.1771.
104. Zhang J, et al. The interferon-stimulated exosomal hACE2 potently inhibits SARS-CoV-2 replication through competitively blocking the virus entry. Signal Transduct Target Ther. 2021a;6(1):1–11. https://doi.org/10.1038/s41392-021-00604-5.
105. Zhang Q, et al. Angiotensin-converting enzyme 2–containing small extracellular vesicles and exomeres bind the severe acute respiratory syndrome coronavirus 2 spike protein. Gastroenterology. 2021b;160(3):958–61. https://doi.org/10.1053/j.gastro.2020.09.042.
106. Zhang X, et al. Effects of mesenchymal stem cells and their exosomes on the healing of large and refractory macular holes. Graefes Arch Clin Exp Ophthalmol. 2018;256:2041–52. https://doi.org/10.1007/s00417-018-4097-3.
107. Zhou T, et al. miR-204-containing exosomes ameliorate GVHD-associated dry eye disease. Sci Adv. 2022;8(2):eabj9617. https://doi.org/10.1126/sciadv.abj9617.
108. Zhou Y, et al. Exosomes released by human umbilical cord mesenchymal stem cells protect against cisplatin-induced renal oxidative stress and apoptosis in vivo and in vitro. Stem Cell Res Ther. 2013;4(2):34. https://doi.org/10.1186/scrt194.
109. Zhu Y-G, et al. Human mesenchymal stem cell microvesicles for treatment of Escherichia coli endotoxin-induced acute lung injury in mice. Stem Cells (Dayton, Ohio). 2014;32(1):116–25. https://doi.org/10.1002/stem.1504.

Chapter 8
Potentiality of Stem Cell–Derived Extracellular Vesicles in the Setting of Metabolic Dysfunction–Associated Steatotic Liver Disease (MASLD)

Angelo Armandi, Gian Paolo Caviglia, and Chiara Rosso

Abbreviations

ChREBP	carbohydrate response element-binding protein
ECM	xtracellular matrix
EVs	extracellular vesicles
FFAs	free fatty acids
DNL	de novo lipogenesis
HLSCs	human liver stem cells
HSC	hepatic stellate cells
IL-6	interleukin-6
iPSC	induced pluripotent stem cells
IR	insulin resistance
IRS	Insulin receptor substrate
JNK	cJun NH2-terminal kinase 1
KC	Kuppfer cells
MCSF	macrophage-colony stimulating factor
mmp-9	matrix metalloproteinase 9
MSCs	Mesenchymal stem cells
MASLD	metabolic dysfunction–associated steatotic liver disease
MASH	metabolic dysfunction–associated steatohepatitis
PPAR	pan-peroxisome proliferator-activated receptor
RIPK	receptor-interacting protein kinase
ROS	reactive oxygen species
SFA	Saturated fatty acids
SIP	sphingosine-1-phosphate

A. Armandi · G. P. Caviglia · C. Rosso (✉)
Department of Medical Sciences, University of Turin, Turin, Italy
e-mail: angelo.armandi@unito.it; gianpaolo.caviglia@unito.it; chiara.rosso@unito.it

P. V. Pham, S. Fagoonee (eds.), *Stem Cell-Derived Exosomes and Microvesicles*, Stem Cells in Clinical Applications, https://doi.org/10.1007/978-3-031-90698-5_8

SOCS	suppressor of cytokine signaling
SREBP-1	sterol regulatory element-binding protein-1
T2DM	type 2 diabetes mellitus
TIMP-1	tissue inhibitor of matrix metalloproteinase 1
TLR9	toll-like receptor 9

1 Introduction

Among all forms of liver disease, metabolic dysfunction–associated steatotic liver (MASLD) represents the most unpredictable entity. Its global prevalence is estimated to be 25% among the adult population, growing in parallel to the widespread burden of metabolic syndrome. In fact, MASLD prevalence is over 90% among obese individuals and can reach 70% in patients with type 2 diabetes mellitus (T2DM). The prevalence of metabolic dysfunction–associated steatohepatitis (MASH), the progressive inflammatory form of liver injury, is less clear, because its diagnosis requires a liver biopsy, which is not commonly available. However, MASH prevalence is estimated to be 5–10% in the general population, but may reach 50% in diabetic and obese individuals with MASLD [1, 2]. Most importantly, MASLD prevalence has been modeled to increase by 30% in 2030, with relevant economic and disease burden [3].

The obesogenic environment is responsible for the onset and progression of MASLD, including sedentary lifestyle, wrong food eating habits, food misbehaviors, and the paradigms of the Western diet that promote insulin resistance. However, the multiple environmental factors have a diverse impact on the liver, in the balance with genetic susceptibility and the individual response to treatment strategies. As a result, subcontinuous bouts of inflammation are responsible for the high inter-individual heterogeneity, in addition to the parallel enhancement of fibrogenesis [4]. However, the deposition of scar tissue is not a linear process, due to the relevant dynamicity of the collagen metabolism, which may induce either regression or progression according to the severity of the inflammatory hits (Fig. 8.1). Over decades, this process becomes more severe and the most relevant determinant of advanced liver disease and portal hypertension. In fact, the deposition of excessive scar tissue can induce pro-inflammatory stimuli and a modification of the endothelium of liver sinusoids, leading to a process known as "capillarization" which is the very first histological stigmata of portal hypertension. This evidence has been confirmed across several in vitro clinical studies. In fact, the progressive deposition of scar tissue inside the liver parenchyma has been widely recognized as the most relevant prognostic factor in the MASLD population. The incidence of overall and liver-related mortality is high in advanced fibrosis, with a stepwise progression across the stages of fibrosis [5]. In lower fibrosis stages, mortality is mainly driven by cardiovascular outcomes and extra-hepatic cancers, mainly due to the concomitant burden of multiple metabolic cofactors. On the contrary, the burden of liver-related diseases becomes significant for higher stages of fibrosis.

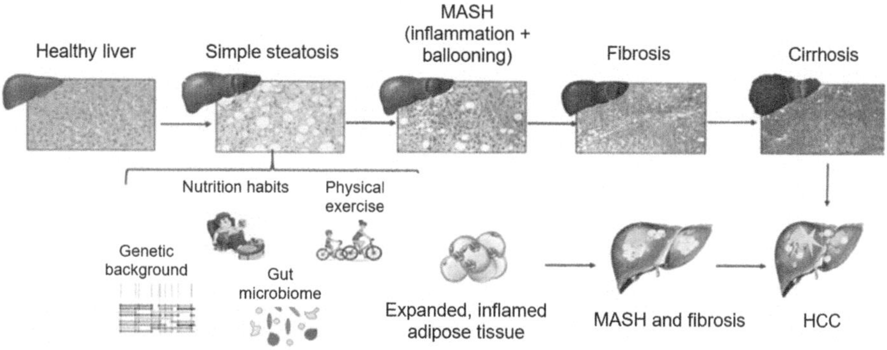

Fig. 8.1 Progression of liver damage in metabolic dysfunction–associated steatotic liver disease Metabolic dysfunction–associated steatotic liver disease (MASLD) represents a wide spectrum of liver damage ranging from isolated hepatic triglyceride accumulation (simple steatosis) to necro-inflammation with or without fibrosis (MASH, metabolic dysfunction–associated steatohepatitis) and ultimately progressing to cirrhosis and potentially hepatocellular carcinoma in the absence of excessive alcohol consumption. MASLD is strongly associated with metabolic syndrome including type 2 diabetes mellitus (T2DM) and dyslipidemia. The increased sedentary lifestyles and changing in dietary patterns contribute to the increase of MASLD prevalence together with the rapid increase of obesity and T2DM. The expanded and inflamed adipose tissue releases pro-inflammatory factors that enhance inflammation and fibrogenesis in the liver

As a result, the identification of patients with significant fibrosis is of utmost importance in clinical settings. In addition, the joint presence of MASH and liver fibrosis (namely fibrosing MASH) has been recognized as an aggressive phenotype of liver injury that demands careful recognition and management.

At present, MASLD treatment relies on lifestyle interventions aiming at weight loss. Multiple dietary patterns have been suggested for the amelioration of liver fat contents, with conflicting results. Overall, the Mediterranean diet has shown optimal results in the improvement of cardiovascular outcomes and systemic metabolism that are closely linked to MASLD [6]. However, food accessibility and geographical and ethnic variabilities limit the success of a unique food intervention strategy. Nonetheless, a weight loss of more than 7% of the initial is linked to a histological improvement of MASH, while a weight loss of more than 10% is associated with an improvement in fibrosis stages [7]. These results are however hindered by lack of compliance and the inability to keep the lifestyle changes over time. Moreover, only one medication, resmetirom, has so far achieved conditional approval by the Food and Drug Administration for the treatment of fibrosing MASH, which is the eligible population for drug treatment. Currently, other targets are being investigated in phase 3 studies for fibrosing MASH: semaglutide, which is a glucagon-like 1-receptor agonist, and lanifibranor, which is a span-peroxisome proliferator-activated receptor (PPAR) agonist. More recently, fibroblast growth factor 21 agonists have shown promising antifibrotic properties and are being investigated in phase 3 studies. Given the multiple metabolic pathways involved in the disease mechanisms, it is unlikely that a single medication will provide a comprehensive improvement. Furthermore, individuals with MASLD without the common metabolic features or with relevant genetic susceptibility should undergo specific treatment approaches that go beyond mere weight loss. Overall, addressing liver fibrosis remains a major challenge and an urgent unmet need.

2 Pathogenesis of MASH

The onset and progression of MASLD is intricate and multifactorial. Along time, different hypotheses have been formulated starting from the "two hit" hypothesis [8]. Accordingly, the first "hit" is the accumulation of triglycerides into the hepatocytes due to insulin resistance (IR) that leads to hepatic steatosis development. Steatosis, in turn, increases the susceptibility of the liver to the second "hit" promoting the generation of free radicals with subsequent release of pro-inflammatory cytokines, activation of inflammatory pathways, and fibrogenesis [9]. Animal models of obesity, characterized by intrahepatic lipids accumulation, showed that a second hit is necessary to promote both inflammation and fibrosis [10].

Over the years, scientific research has increasingly deepened the molecular mechanisms involved in the progression of liver damage in patients with MASLD leading to the awareness that multiple factors act synergistically in genetically predisposed individuals and are responsible for the development and progression of disease. It had become clear that the "two hit" theory had to be revised taking into consideration a "multiple hit" hypothesis [11].

2.1 *Insulin Resistance, Inflammation, and Lipotoxicity*

The most important pathogenic mechanism involved in the development of hepatic steatosis and its progression to MASH is IR, which is defined as a less-than-expected response to insulin stimulation by tissues. In the adipose tissue, IR is characterized by an impaired suppression of lipolysis due to the inhibition of the lipase with the consequent massive release of free fatty acids from the adipocytes reaching the liver and other organs. In the liver, IR leads to an impaired suppression of endogenous glucose production with the consequent release of glucose contributing to hyperglycemia and compensatory hyperinsulinemia. In the skeletal muscle, IR causes an impaired uptake of glucose by the myocytes contributing to increased blood glucose concentration which worsens hyperinsulinemia (Fig. 8.2).

Obesity and genetic predisposition may contribute to IR development, which is a common feature in patients with MASLD. Indeed, IR is more prevalent in patients with MASH compared to those with simple steatosis and in obese compared to lean individuals [12]. In MASLD patients, both genetic and environmental factors affect insulin signaling pathways perpetuating both inflammation and IR. For example, a

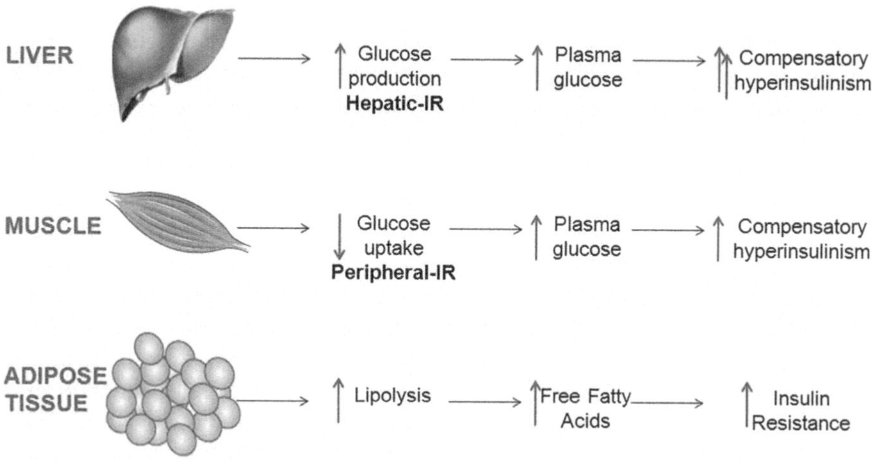

Fig. 8.2 Sites and mechanisms of insulin resistance
In patients with MASLD, insulin resistance (IR) is considered the main pathogenic mechanism involved in the progression of liver damage. Specifically, IR in the adipose tissue is tightly associated with liver fibrosis; the impaired suppression of lipolysis leads to a massive release of free fatty acids (FFAs) that from the adipocytes reach the liver contributing to the development of steatosis. The increased overload of FFAs in the hepatocytes enhances lipotoxicity and inflammatory processes which in turn promote the activation of hepatic stellate cells, and thus fibrogenesis

diet rich in fat induces IR in the liver through the regulatory protein cJun NH2-terminal kinase 1 (JNK1). Knock-out mice in which the JNK1 gene was selectively ablated in the adipose tissue showed an increased hepatic IR. In addition, JNK1 can stimulate the release of the pro-inflammatory interleukin (IL)-6 by the adipose tissue increasing the expression of liver suppressor of cytokine signaling (SOCS) 3 that in turn induces IR [13]. The SOCS proteins are upregulated in obese patients and are involved in the pathogenesis of T2DM [14]. Insulin receptor substrate (IRS) proteins are also involved in the development of IR. Specifically, serine phosphorylation of IRS-1 is increased in insulin-resistant conditions and the entity of serine phosphorylation negatively regulates IRS signaling which in turn contributes to the pathogenesis of IR [15]. Protein-kinase B (Akt) 2 deficient mice are characterized by IR and develop T2DM as a consequence of glucose metabolism derangement [16]. In addition, genetic polymorphisms in the Akt2 gene have been found and are associated with IR and metabolic syndrome [17, 18].

The accumulation of free fatty acids (FFAs) in the liver is the main mechanism leading to the development of hepatic steatosis, accounting for 60%; 30% of triglycerides comes from de novo lipogenesis (DNL) while only 10% derives from diet. In normal conditions, the accumulation of fat in the form of triglycerides represents a protective mechanism against lipotoxicity [19]; but when the excess of FFAs in the liver exceeds the ability to synthesize triglycerides, lipotoxicity phenomena may originate. Lipotoxicity plays a relevant role in the pathogenesis of MASH; it stimulates an inflammatory response which is associated with the activation of macrophage and pro-apoptotic signaling [20]. Lipotoxicity induces lysosomal dysfunction in the hepatocytes favoring the release of exosomes that in turn act on macrophages addressing their differentiation toward M1 polarization, a pro-inflammatory phenotype. In vitro studies showed that palmitic acid, a saturated acid with lipotoxic activity, promotes the release of exosomes reach in miR-192-5p from the hepatocytes which in turn play a critical role in the activation of M1 macrophages [21]. This evidence demonstrates that reducing lipotoxicity and inhibiting the release of exosomes, or targeting miR-192-5p, may ameliorate liver inflammation.

DNL, the enzymatic pathway through which dietary carbohydrates are converted into fat, is threefold higher in patients with MASLD compared to healthy subjects [22], and fivefold higher in hyperinsulinemic patients undergoing a diet rich in carbohydrate compared to normoinsulinemic subjects [23]. Glucose is the main substrate for DNL while fructose is able to promote DNL independently from hyperinsulinemia [24, 25]. Molecular mechanisms leading to the activation of DNL depend on the activation of several transcription factors such as sterol regulatory element-binding protein-1 (SREBP-1), carbohydrate response element-binding protein (ChREBP), and PPAR-gamma [26]. Specifically, the dysregulation of SREBP promotes hepatic fat accumulation [27]. ChREBP is activated by glucose and is able to modulate both FFAs and triglycerides metabolism and enhances DNL pathway. In IR condition, the downregulation of IRS-2 leads to an overexpression of SREBP-1 which in turn contributes to the increase in DNL [28].

Another mechanism contributing to the increased lipotoxicity in the setting of MASLD is the inhibition of carnitine palmitoyltransferase and the consequent reduction of mitochondria b-oxidation of FFAs [29]. Beyond all these mechanisms, it is clear that not only the quantity of accumulated lipids but also the type of lipids accumulated in the hepatocytes impacts on liver disease progression contributing to inflammation and fibrosis [30]. The excess of FFAs that are not esterified in more complex lipidic species may lead to the formation of reactive oxygen species (ROS) through endoplasmic reticulum stress, mitochondrial dysfunction, and oxidative stress. As a response, a pro-inflammatory and pro-fibrogenic pathway is activated promoting the progression of liver disease [31]. Saturated fatty acids (SFA) are able to drive macrophage differentiation towards the pro-inflammatory phenotype M1. As previously reported, these macrophages release pro-inflammatory cytokines and enhance IR. Conversely, monounsaturated and polyunsaturated fatty acids (MUFA and PUFA) seem to improve glucose metabolism by reducing IR, thus improving hepatic steatosis [32]. Despite PUFA seem to have a beneficial effect in the context of MASLD progression, there is a difference between n-6 PUFA (including arachidonic acid) and n-3 PUFA (including eicosapentaenoic and docosahexaenoic acids). Several studies investigated the possible positive effect of n-3 PUFA; they seem to affect liver function, oxidative stress and endothelial function through their anti-inflammatory properties. Conversely, a pro-thrombotic and pro-inflammatory effect of n-6 PUFA has been described [33]. Interestingly, the increased n-6 PUFA/n-3 PUFA ratio seems to be associated with a higher risk of obesity and MASLD suggesting the importance of n-3 PUFA supplementation with diet [34–37].

The excess of FFAs leads to the increased synthesis of ceramides, another important source of toxic metabolites [38]. Ceramides are involved in the disruption of insulin signaling pathways, enhance oxidative stress, increase the expression of pro-inflammatory cytokines promoting inflammation, and seem to activate hepatic stellate cells (HSC) thus promoting fibrogenesis [39]. Endoplasmic reticulum stress mediates hepatocytes' EVs release; these EVs contain lipotoxic metabolites such as ceramides and sphingosine-1-phosphate (S1P) that again induce macrophage activation and their chemotaxis [40].

2.2 *Physiopathology of Hepatic Fibrosis*

Liver fibrosis is considered a reversible wound-healing response which is characterized by improved deposition of extracellular matrix (ECM). Hepatocellular damage leads to macrophage activation which in turn acts on HSC thus promoting collagen accumulation. In case of sustained injury, chronic inflammation and collagen deposition persist leading to the progressive substitution of liver parenchyma with non-functional scar tissue.

The main ECM components are proteoglycans (i.e., glycosaminoglycans, heparan sulfate, hyaluronic acid, decorin) and fibrous proteins such as elastin and collagen (structural proteins), and adhesion proteins such as laminin, fibronectin, and entactin. ECM can be classified as follows: (1) the interstitial matrix, which is composed of several types of collagens (I, III, V) plus fibronectin and proteoglycans; (2) the basement membrane, which is characterized by the presence of type IV collagen and laminins and forms an adhesion interface between parenchymal cells and the more external ECM. All these components can interact with different cytokines and growth factors promoting myofibroblasts activation, chemotaxis and proliferation, angiogenesis, epithelial cell proliferation, and survival [41].

The main cells involved in the fibrogenesis process are hepatic stellate cells (HSC); these cells receive paracrine stimuli from surrounding cells such as Kuppfer cells (KC), which enhance fibrogenesis in the liver. Nevertheless, the interaction between pro-inflammatory and pro-fibrogenic factors in promoting liver damage remains to be clarified. New evidence has shown that the majority of the stimuli are transmitted via exosomes which increase in pathological conditions perpetuating inflammation and driving fibrogenesis [42]. One of the most relevant mechanisms involved in the activation of HSC is autophagy, a process by which both organelles and macromolecules are degraded through the lysosomal pathway [43]. Autophagy, a highly regulated process important for the maintenance of liver homeostasis, increases under stress conditions [44] and its dysfunctions are associated with the progression of liver disease [45]. Specifically, autophagy seems to modulate hepatic fibrogenesis through pro- and antifibrogenic factors involving the Akt/mammalian target of rapamycin (PI3K/Akt/mTOR) [46]. Another important mechanism involved in the progression of MASH and liver fibrosis is necroptosis, the programmed form of necrosis [47, 48]. Necroptosis is mediated by the receptor-interacting protein kinase (RIPK)-1 and RIPK-3; specifically, a mouse model of MASH in which RIPK-3 was genetically inhibited showed an improvement of liver fibrosis [49].

Another mechanism involved in the process of fibrogenesis is mediated by the activation of inflammasome NLRP3 which contributes to the massive release of ROS that in turn drive the production of extracellular matrix (type I collagen and a-SMA) in HSCs [50]. Therapeutic strategies against inflammation, oxidative stress, and apoptosis may improve NAFLD progression.

2.3 Extracellular Vesicles and Their Role in MASLD Progression

EVs are small membrane particles, which are released actively from the cells. According to their size, EVs are grouped into exosomes (30–100 nm), microvesicles (100–1000 nm), and apoptotic bodies (from 50 to 500 nm); according to their cargo (DNA, RNA, microRNA, or long noncoding RNA), can modulate gene expression and molecular pathways in target cells [51]. The release of EVs increases during pathological conditions such as viral infection, tumors, and metabolic diseases. For example, in subjects with MAFLD, EVs can be released by the adipose tissue and can modulate endocrine effects in several cell populations such as KC, HSCs, and macrophages enhancing inflammation and promoting the development of fibrosis [52] (Fig. 8.3). In 2009, Deng et al. described for the first time a role of adipose tissue-derived EVs in the modulation of insulin resistance in a mouse model of obesity [53]; briefly, EVs from adipocytes of obese and high-fat-fed mice are able to activate macrophage increasing the release of pro-inflammatory factors such as tumor necrosis factor (TNF)-a, interleukin-6 (IL-6), and macrophage-colony stimulating factor (MCSF). Moreover, cultured media isolated from the activated macrophages disrupted glucose metabolism favoring the development of IR in the skeletal muscle. In another study, EVs isolated from adipose tissue macrophages of lean mice injected into obese mice were able to normalize glucose metabolism improving insulin sensitivity [54]. Notably, the obese adipocytes EVs are able to target HSCs and hepatocytes inducing the expression of matrix metalloproteinase 9 (MMP-9) and tissue inhibitor of matrix metalloproteinase 1 (TIMP-1), thus promoting fibrogenesis [55]. Besides adipose tissue, damaged hepatocytes also release EVs as a consequence of the excessive lipids accumulation; these EVs contain several pro-inflammatory factors which promote liver disease progression acting on both KCs and HSCs [56]. Hepatocyte-derived EVs from mice models of NASH contain high levels of mitochondrial DNA which can activate the toll-like receptor 9 (TLR9) resulting in the activation of NF-kB pro-inflammatory pathway in macrophages thus perpetuating liver disease [57, 58].

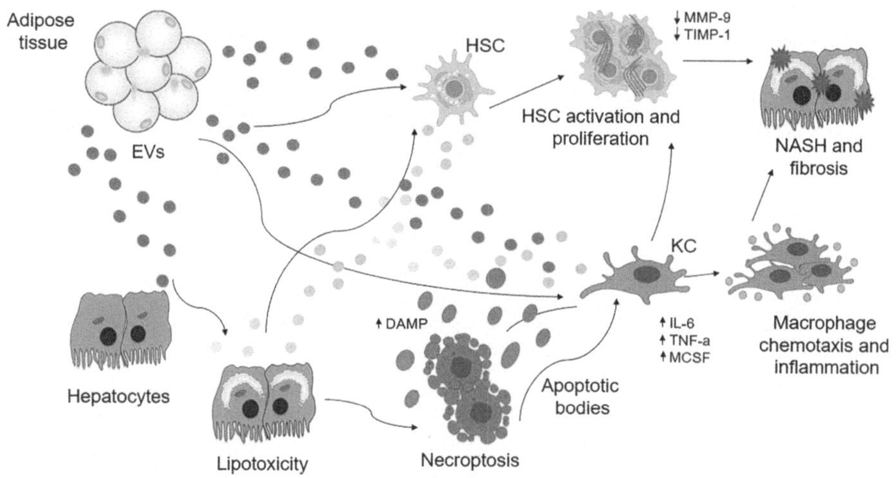

Fig. 8.3 Extracellular vesicles and their role in the progression of MASLD
The release of extracellular vesicles (EVs) increases during pathological conditions such as viral infection, tumors, and metabolic diseases. In the context of MASLD, EVs can be released by the adipose tissue and can modulate endocrine effects in several cell populations such as Kuppfer Cells (KCs), hepatic stellate cells (HSCs), and macrophages enhancing inflammation and promoting the development of fibrosis
Abbreviations: *DAMP* damage associated molecular pattern, *EVs* extravesicles, *KC* Kuppfer cells, *HSC* hepatic stellate cells, *IL-6* interleukin-6, *MCSF* macrophage colony-stimulating factor, *MMP-9* matrix metalloproteinase 9, *MASH* metabolic dysfunction-associated steatohepatitis, *TIMP-1* tissue inhibitor of metalloproteinase, *TNF-a* tumor necrosis factor alpha

3 Mesenchymal Stem Cells as Treatment Option in the Setting of MASLD

Mesenchymal stem cells (MSCs) are being studied in different liver diseases including MASLD. In the complex scenario of MASLD pathogenesis, the MSCs-based therapy may introduce a potential alternative to classical regimens and lead to more tailored approaches. However, the evolving landscape of MSCs has yielded some drawbacks. High costs, time expenditure linked to generation, handling, and phenotype stability have been reported. Moreover, ectopic tissue formation, rejection, or infusional toxicity represents major limitations. In this context, the exploitation of MSC-secreted EVs can compensate for these limitations, as they are smaller than cells and easier to produce. Interestingly, MSCs can differentiate into different cell types and, from a clinical point of view, they are of great interest for their ability to secrete EVs which in turn may transport chemokines, cytokines and other factors able to modulate specific recipient cells.

In particular, exosomes derived from diverse types of MSCs have provided positive evidence for liver disease management in preclinical studies conducted on animal models (Table 8.1). For instance, exosomes derived from MSCs obtained from

Table 8.1 Overview of preclinical studies conducted on murine models using diverse sources of exosomes derived from mesenchymal stem cell (MSC) to address liver fibrosis, liver injury, and metabolic cofactors

Study	Source of MSC-derived EV	Target mechanism	Therapeutic effect
Li et al. [59]	Human Wharton's jelly (direct effect of MSCs)	Suppression of epithelial-mesenchymal transition Suppression of TGFβ1/Smad pathway Reduction in type I and III collagen deposition	Reduced liver fibrosis
Hyun et al. [60]	Chorionic plate (direct effect of MSCs)	Inactivation of Hedgehog signaling by exosomal miR-125b	Reduced liver fibrosis
Lou et al. [61]	Adipose tissue	Suppression of collagen maturation and HSC activation by exosomal miR-122	Reduced liver fibrosis Reduced serum levels of hyaluronic acid and procollagen III N-peptide
Tan et al. [62]	Human embryonic MSCs	Upregulation of cyclin D1 Upregulation of anti-apoptotic Bcl-xL gene	Hepatoprotective effect in acute liver failure
Tamura et al. [63]	Bone marrow	Reduced pro-inflammatory cytokines Reduced apoptosis Enhanced anti-inflammatory cytokines Increased T-reg cells	Reduced serum transaminases Improved liver injury
Yan et al. [64]	Human Wharton's jelly	Improvement of oxidative stress and reduction in oxygen reactive products by glutathione peroxidase 1	Treated acute liver injury
Haga et al. [65]	Bone marrow	Modulation of cytokine responses, reduction in inflammatory processes and apoptosis by exosomal Y-RNA-1	Reduced liver injury
Sun et al. [68]	Human umbilical cord	Restored phosphorylation in insulin receptors substrates Enhanced translocation of GLUT4 in skeletal muscle Increased glycogen storage in the liver Inhibition of pancreatic β cell apoptosis	Reversal of peripheral IR in T2DM
Povero et al. [70]	Human-induced pluripotent stem cell	Decrease in alpha-smooth muscle actin, fibronectin and tissue inhibitor of metalloproteinases-1 Modulation of chemotaxis and proliferation Increased expression of miR-92a-3p	Reduced liver fibrosis
Bruno et al. [66]	Human liver stem cells	Modulation of gene expression: genes involved in ECM remodeling, TGF signaling pathway, and inflammatory cytokines	Reduce liver fibrosis and inflammation

human Wharton's jelly and transplanted into a murine model of carbon tetrachloride-induced liver injury, has shown to improve liver fibrosis by inactivating transforming growth factor β1/Smad pathway and subsequent inhibition of collagen I/III production. A parallel improvement in serum transaminase levels was observed. In addition, a further antifibrotic effect of MSC-derived exosomes was exerted upon the inhibition of epithelial-to-mesenchymal transition as shown by the decrease in N-cadherin and vimentin-positive cells [59]. Another study exploited the use of human chorionic plate-derived MSCs, which proved to express and release miR-125b. In this study, these human MSCs expressing miR-125b had the ability to act upon murine HSCs, inhibiting liver fibrosis by inactivation of the Hedgehog signaling pathway [60]. Interestingly, the use of adipose tissue–derived MSCs has provided similar results. These MSCs express mirR-122 and convey it to HSCs, where it modulates cell proliferation and collagen synthesis through target genes. This epigenetic modulation improved the circulating levels of procollagen III N-peptide and hyaluronic acid [61], which are among the most promising biomarkers in the noninvasive identification of liver fibrosis in the MASLD population. The dialogue between mirR-122 expressing MSCs and HSCs was mediated by exosomes, further encouraging the use of EVs in the modulation of liver fibrosis. Moreover, this approach would overcome the limited feasibility of the straight use of MSCs.

Positive results have also been obtained in the alleviation of liver inflammation. Human embryonic MSCs exert hepatoprotective effects upon the anti-apoptotic Bcl-xL gene, enhancing tissue regeneration [62]. In a mouse model of immune-induced liver injury, EVs of MSCs derived from bone marrow could reduce both transaminase and pro-inflammatory cytokines and, interestingly, upregulate the activity of T-reg cells [63]. The evidence of a modulation of the inflammasome and immune system may suggest the most promising role for MSCs in the hepatology field. EVs obtained from MSCs Wharton's jelly-derived MSCs were also exploited for the carriage of glutathione peroxidase-1 upon a mouse model of acute liver injury, where they proved to clear oxygen reactive products and improve oxidative stress [64]. As mentioned above, the onset of oxidative stress inside the hepatocytes is a major source of liver injury in the transition from simple steatosis to MASH and favors the progressiveness of liver disease. Shuttling enzymes involved in the modulation of oxidative stress may represent a great opportunity to heal the injury at the very beginning. Another study showed a potential application of exosomal Y-RNA derived from bone marrow MSCs. Y-RNA are highly-conserved, mid-sized noncoding extracellular RNAs that are more representative than miRNA of the small non-coding RNA molecules contained in exosomes. In particular, Y-RNA-1 proved to modulate cytokine expression, exert anti-apoptotic effect, and improve survival in a murine model of tumor necrosis factor α-induced acute liver failure [65].

Remarkable results have been observed in the murine model of MASH treated with EVs from human liver stem cells (HSLCs) [66], a stem cell population derived from human adult liver cells. This type of stem cells shares with MSCs the ability to differentiate and similar immunomodulatory properties but shows a specific hepatic commitment. A recent study by Bruno et al. investigated whether HLSC-EVs could influence the progression of MASH in immunocompromised mice; overall, the authors observed that HLSC-EVs were able to attenuate liver fibrosis and inflammation at both morphological and molecular levels [67]. In particular, it has been observed that EVs reduced ALT values and improved albumin synthesis in comparison to MASH mice treated with vesicle hicle only. Furthermore, treatment with EVs induced the expression of genes encoding ECM remodeling enzymes, transforming growth factor beta-1 signaling pathway, as well as inflammatory cytokines involved in the onset and progression of liver fibrosis. Interestingly, no safety issues were pointed out; after systemic administration, EVs derived from HLSCs localized in the liver, with a peculiar tropism for fibrotic areas.

The evaluation of MASLD cannot ignore the multidimensional features of metabolic syndrome. Indeed, the application of MSCs-derived exosomes has shown positive effects in the management of T2DM. Since T2DM is a major metabolic burden in the MASLD population, its proper management induces a concomitant alleviation of liver disease. In one study, the use of MSCs from a human umbilical cord in a mouse model of diabetes led to a reversion in peripheral IR through restoration of phosphorylation in the insulin receptor substrates, a translocation of GLUT-4 in the skeletal muscle, and increased storage of glycogen in the liver [68]. Similarly, potential positive effects may derive from the use of MSCs addressing specific features of metabolic syndrome impacting the liver (dyslipidemia, arterial hypertension, visceral obesity, chronic kidney disease).

Apart from MSCs, another promising therapeutic approach is represented by induced pluripotent stem cells (iPSCs), which have already been shown to be able to downregulate the expression of pro-inflammatory and pro-fibrogenic genes [69]. Though iPSCs have greater differentiation capacity than multipotent stem cells, several safety issues, including immunogenic and oncogenic potential, strongly limited their use as therapeutics; however, the possibility to generate and harvest ex vivo stem cell–derived EVs could overcome this issue. A recent study explored whether iPSC-EVs may reduce or even revert liver fibrosis in vitro and in the animal model [70]. iPSC-EVs were produced in vitro; ultrastructure of EVs was assessed by transmission electron microscopy and confirmed by specific EV markers, such as tetraspanins CD63 and CD81. In vitro, it has been demonstrated that iPSC-EVs were internalized by HSCs, and were able to modulate pro-fibrogenic genes and reduce the expression of cell proliferation markers. Furthermore, HSCs primed with transforming growth factor (TGF)-β and exposed to iPSC-EVs showed a decreased chemotaxis compared to TGF-β primed HSCs exposed to EV-free supernatant. In vivo, following intravenous administration in C57BL/6 mice, iPSCs showed a specific tropism for the liver; in the CCl_4-induced murine model of liver damage, administration of EVs induced a reduction of fibrosis up to 30%. By transcriptomic analysis, activated HSCs exposed to iPSC-EVs allowed the identification of 30 genes specifically downregulated by EV exposure. Taken together, these results suggest that iPSC-EVs can modulate HSCs inducing an antifibrotic effect without safety concerns, thus representing an additional potential novel antifibrotic biologics for the treatment of liver damage in patients with MASLD.

4 Conclusion

The preclinical studies encourage the potential use of MSCs-derived EVs in human MASLD. The diverse source of MSCs allows for a multipotent cellularity with specific modulation strategies. The epigenetic regulation induced by miRNA or Y-RNA, or the shuttling of crucial enzymes involved in pathogenic crosslinks, represents novel approaches to the management of complex diseases. Reversal of liver fibrosis is on top of the ongoing research agenda, and multiple drugs are being evaluated for the alleviation of liver inflammation in MASLD. The multiple disease pathways account for the high heterogeneity in MASLD individuals and this is frequently observed in clinical practice. As a consequence, the extremely tailored approach that is offered by MSCs-derived exosomes can help elucidate the physiopathology of MASLD, which is still incompletely understood, as well as act on specific disease patterns that would increase the treatment response rate, which is still unsatisfactory.

Competing Interests The authors declare that they have no competing interests.

References

1. Younossi ZM, Koenig AB, Abdelatif D, Fazel Y, Henry L, Wymer M. Global epidemiology of nonalcoholic fatty liver disease-Meta-analytic assessment of prevalence, incidence, and outcomes. Hepatology. 2016;64(1):73–84.
2. Bhala N, Younes R, Bugianesi E. Epidemiology and natural history of patients with NAFLD. Curr Pharm Des. 2013;19(29):5169–76.
3. Estes C, Anstee QM, Arias-Loste MT, Bantel H, Bellentani S, Caballeria J, et al. Modeling NAFLD disease burden in China, France, Germany, Italy, Japan, Spain, United Kingdom, and United States for the period 2016-2030. J Hepatol. 2018;69(4):896–904.
4. Schuppan D, Surabattula R, Wang XY. Determinants of fibrosis progression and regression in NASH. J Hepatol. 2018;68(2):238–50.
5. Dulai PS, Singh S, Patel J, Soni M, Prokop LJ, Younossi Z, et al. Increased risk of mortality by fibrosis stage in nonalcoholic fatty liver disease: systematic review and meta-analysis. Hepatology. 2017;65(5):1557–65.
6. Kastorini CM, Milionis HJ, Esposito K, Giugliano D, Goudevenos JA, Panagiotakos DB. The effect of Mediterranean diet on metabolic syndrome and its components: a meta-analysis of 50 studies and 534,906 individuals. J Am Coll Cardiol. 2011;57(11):1299–313.
7. Vilar-Gomez E, Martinez-Perez Y, Calzadilla-Bertot L, Torres-Gonzalez A, Gra-Oramas B, Gonzalez-Fabian L, et al. Weight loss through lifestyle modification significantly reduces features of nonalcoholic steatohepatitis. Gastroenterology. 2015;149(2):367–78.
8. Day CP, James OFW. Steatohepatitis: a tale of two hits? Gastroenterology. 1998;114:842–5.
9. Peverill W, PowellLW SR. Evolving concepts in the pathogenesis of NASH: beyond steatosis and inflammation. Int J Mol Sci. 2014;15:8591–638.
10. Leamy AK, Egnatchik RA, Young JD. Molecular mechanisms and the role of saturated fatty acids in the progression of non-alcoholic fatty liver disease. Prog Lipid Res. 2013;52:165–74.
11. Buzzetti E, Pinzani M, Tsochatziz A. The multiple-hit pathogenesis of non-alcoholic fatty liver disease (NAFLD). Metab Clin Exp. 2016;65:1038–48.

12. Pagano G, Pacini G, Musso G, Gambino R, Mecca F, Depretis N, et al. Nonalcoholic steatohepatitis, insulin resistance, and metabolic syndrome: further evidence for an etiologic association. Hepatology. 2002;35:367–72.
13. Sabio G, Das M, Mora A, Zhang Z, Jun JY, Ko HJ, et al. A stress signaling pathway in adipose tissue regulates hepatic insulin resistance. Science. 2008;322:1539–43.
14. Emanuelli B, Peraldi P, Filloux C, Chavey C, et al. SOCS-3 inhibits insulin signaling and is up-regulated in response to tumor necrosis factor-α in the adipose tissue of obese mice. J Biol Chem. 2001;276:47944–9.
15. Taniguchi CM, Emanuelli B, Khan CR. Critical nodes in signalling pathways: insights into insulin action. Nat Rev Mol Cell Biol. 2006;7:85–96.
16. Cho H, Mu J, Kim JK, Thorvaldsen JL, Chu Q, Crenshaw EB, et al. Insulin resistance and a diabetes mellitus-like syndrome in mice lacking the protein kinase Akt2 (PKBB). Science. 2001;292:1728–31.
17. George S, Rochford JJ, Wolfrum C, Gray SL, Schinner S, Wilson JC, et al. A family with severe insulin resistance and diabetes due to a mutation in AKT2. Science. 2004;304:1325–8.
18. Tassi V, Coco A, Padovano L, De Bonis C, De Cosmo S, Trischitta V. Akt2 gene common allelic variants in insulin resistance and the metabolic syndrome. Nutr Metab Cardiovasc Dis. 2008;18:263–70.
19. Koliwad SK, Streeper RS, Monetti M, Cornelissen I, Chan L, Terayama K, et al. DGAT1-dependent triacylglycerol storage by macrophages protects mice from diet-induced insulin resistance and inflammation. J Clin Invest. 2010;120:756–67.
20. Hirsova P, Ibrahim SH, Gores GJ, Malhi H. Lipotoxic lethal and sublethal stress signaling in hepatocytes: relevance to NASH pathogenesis. J Lipid Res. 2016;57:1758–70.
21. Liu XL, Pan Q, Cao HX, Xin FZ, Zhao ZH, Yang RX, et al. Lipotoxic hepatocyte-derived exosomal MicroRNA 192-5p activates macrophages through Rictor/Akt/Forkhead box transcription factor O1 signaling in nonalcoholic fatty liver disease. Hepatology. 2020;72:4.
22. Lambert JE, Ramos-Roman MA, Browning JD, Parks EJ. Increased de novo lipogenesis is a distinct characteristic of individuals with nonalcoholic fatty liver disease. Gastroenterology. 2014;146:726–35.
23. Schwarz JM, Linfoot P, Dare D, Aghajanian K. Hepatic de novo lipogenesis in normoinsulinemic and hyperinsulinemic subjects consuming high-fat, low-carbohydrate and low-fat, high-carbohydrate isoenergetic diets. Am J Clin Nutr. 2003;77:43–50.
24. Dashty M. A quick look at biochemistry: carbohydrate metabolism. Clin Biochem. 2013;46:1339–52.
25. Jensen T, Abdelmalek MF, Sullivan S, Nadeau KJ, Green M, Roncal C, et al. Fructose and sugar: a major mediator of non-alcoholic fatty liver disease. J Hepatol. 2018;68:1063–75.
26. George J, Liddle C. Nonalcoholic fatty liver disease: pathogenesis and potential for nuclear receptors as therapeutic targets. Mol Pharm. 2008;5:49–59.
27. Musso G, Gambino R, Cassader M. Cholesterol metabolism and the pathogenesis of non-alcoholic steatohepatitis. Prog Lipid Res. 2013;52:175–91.
28. Stefan N, Kantartzis K, Haring HU. Causes and metabolic consequences of fatty liver. Endocr Rev. 2008;29:939–60.
29. Reddy JK, Rao MS. Lipid metabolism and liver inflammation II. Fatty liver disease and fatty acid oxidation. Am J Physiol Gastrointest Liver Physiol. 2006;290:852–8.
30. Farrell GC, Haczeyni F, Chitturi S. Pathogenesis of NASH: how metabolic complications of overnutrition favour lipotoxicity and pro-inflammatory fatty liver disease. Adv Exp Med Biol. 2018;1061:19–44.
31. Alves-Bezerra M, Cohen DE. Triglyceride metabolism in the liver. Compr Physiol. 2017;8:1–8.
32. Yki-Jarvinen H. Nutritional modulation of non-alcoholic fatty liver disease and insulin resistance. Nutrients. 2015;7:9127–38.
33. Monteiro J, Leslie M, Moghadasian MH, Arendt BM, Allard JP, Ma DW. The role of n-6 and n-3 polyunsaturated fatty acids in the manifestation of the metabolic syndrome in cardiovascular disease and non-alcoholic fatty liver disease. Food Funct. 2014;5:426–35.

34. Patterson E, Wall R, Fitzgerald GF, Ross RP, Stanton C. Health implications of dietary omega-6 polyunsaturated fatty acids. J Nutr Metab. 2012;2012:539426.
35. Simopoulos AP. An increase in the omega-6/omega-3 fatty acid ratio increases the risk for obesity. Nutrients. 2016;8:128.
36. Janczyk W, Socha P, Lebensztejn D, Mazur A, Neuhoff-Murawska J, Matusik P. Omega-3 fatty acids for treatment of non-alcoholic fatty liver disease: design and rationale of randomized controlled trial. BMC Pediatr. 2013;13:85.
37. Nobili V, Alisi A, Musso G, Scorletti E, Calder PC, Byrne CD. Omega-3 fatty acids: mechanisms of benefit and therapeutic effects in pediatric and adult NAFLD. Crit Rev Clin Lab Sci. 2016;53:106–20.
38. Martinez L, Torres S, Baulies A, Alarcon-Vila C, Elena M, Fabriàs C, et al. Myristic acid potentiates acid-induced lipotoxicity and steatohepatitis associated with lipodystrophy by sustaining de novo ceramide synthesis. Oncotarget. 2015;6:41479–96.
39. Hajduch E, Lachhkar F, Ferré P, Foufelle F. Roles of ceramides in non-alcoholic fatty liver disease. JClinMed. 2021;10:792.
40. Liao CY, Song MJ, Gao Y, Mauer AS, Revzin A, Malhi H. Hepatocyte-derived lipotoxic extracellular vesicle sphingosine 1-phosphate induces macrophage chemotaxis. Front Immunol. 2018;9:2980.
41. Karsdal MA, Manon-Jensen T, Genovese F, Kristensen JH, Nielsen MJ, Sand JM, et al. Novel insights into the function and dynamics of extracellular matrix in liver fibrosis. Am J Physiol Gastrointest Liver Physiol. 2015;308(10):G807–30.
42. Mastoridou EM, Goussia AC, Glantzounis GK, Kanavaros P, Charchanti AV. Autophagy and exosomes: cross-regulated pathways playing major roles in hepatic stellate cells activation and liver fibrosis. Phront Physiol. 2022;12:801340.
43. Weiskirchen R, Tacke F. Liver fibrosis: from pathogenesis to novel therapies. Dig Dis. 2016;34:410–22.
44. Parzych KR, Klionsky DJ. An overview of autophagy: morphology, mechanism, and regulation. Antioxid Redox Signal. 2014;20(3):460–73.
45. Wirawan E, Vanden Berghe T, Lippens S, Agostinis P, Vandenabeele P. Autophagy: for better or for worse. Cell Res. 2012;22(1):43–61.
46. Hernandez-Gea V, Ghiassi-Nejad Z, Rozenfeld R, Gordon R, Fiel MI, Yue Z, Czaja M, Friedman S. Autophagy releases lipid that promotes fibrogenesis by activated hepatic stellate cells in mice and in human tissues. Gastroenterology. 2012;142:938–46.
47. Schwabe RF, Luedde T. Apoptosis and necroptosis in the liver: a matter of life and death. Nat Rev Gastroenterol Hepatol. 2018;15:738–52.
48. Koppe C, Verheugd P, Gautheron J, Reisinger F, Kreggenwinkel K, Roderburg C, et al. IkappaB kinasealpha/beta control biliary homeostasis and hepatocarcinogenesis in mice by phosphorylating the cell-death mediator receptor-interacting protein kinase 1. Hepatology. 2016;64:1217–31.
49. Molnár T, Mazlo A, Tslaf V, Szollosi AG, Emri G, Koncz G. Current translational potential and underlying molecular mechanisms of necroptosis. Cell Death Dis. 2019;10:860.
50. Ma Y, Lee G, Heo S, Roh Y. Oxidative stress is a key modulator in the development of nonalcoholic fatty liver disease. Antioxidants. 2021;11:91.
51. Kalluri R, LeBleu VS. The biology, function and medical application of exosomes. Science. 2020;367
52. Dorairaj V, Sulaiman SA, Abu N, Abdul Murad NA. Extracellular vesicles in the development of the non-alcoholic fatty liver disease: an update. Biomol Ther. 2020;10:1494.
53. Deng ZB, Poliakov A, Hardy RW, Clements R, Liu C, Liu Y, et al. Adipose tissue exosome-like vesicles mediate activation of macrophage-induced insulin resistance. Diabetes. 2009;58:2498–505.
54. Ying W, Riopel M, Bandyopadhyay G, Dong Y, Birmingham A, Seo JB, et al. Adipose tissue macrophage-derived exosomal mirnas can modulate in vivo and in vitro insulin sensitivity. Cell. 2017;171:372–84.

55. Koeck ES, Iordanskaia T, Sevilla S, Ferrante SC, Hubal MJ, Freishtat RJ, et al. Adipocytes exosomes induce transforming growth factor beta pathway dysregulation in hepatocytes: a novel paradigm for obesity-related liver disease. J Surg Res. 2014;192:268–75.
56. Szabo G, Momen-Heravi F. Extracellular vesicles in liver disease and potential as biomarkers and therapeutic targets. Nat Rev Gastroenterol Hepatol. 2017;14:455–66.
57. Garcia-Martinez I, Santoro N, Chen Y, Hoque R, Ouyang X, Caprio S, et al. Hepatocytes mitochondrial DNA drives nonalcoholic steatohepatitis by activation of TLR9. J Clin Invest. 2016;126:859–64.
58. Sasai M, Linehan MM, Iwasaki A. Bifurcation of toll-like receptor 9 signaling by adaptor protein 3. Science. 2010;329:1530–4.
59. Li T, Yan Y, Wang B, Qian H, Zhang X, Shen L, et al. Exosomes derived from human umbilical cord mesenchymal stem cells alleviate liver fibrosis. Stem Cells Dev. 2013;15(22):845–54.
60. Hyun J, Wang S, Kim J, Kim GJ, Jung Y. MicroRNA125b-mediated Hedgehog signaling influences liver regeneration by chorionic plate-derived mesenchymal stem cells. Sci Rep. 2015;5:14135.
61. Lou G, Yang Y, Liu F, Ye B, Chen Z, Zheng M, et al. MiR-122 modification enhances the therapeutic efficacy of adipose tissue-derived mesenchymal stem cells against liver fibrosis. J Hematol Oncol. 2015;8:122.
62. Tan CY, Lai RC, Wong W, Dan YY, Lim SK, Ho HK. Mesenchymal stem cell-derived exosomes promote hepatic regeneration in drug-induced liver injury models. Stem Cell Res Ther. 2014;5:76.
63. Tamura R, Uemoto S, Tabata Y. Immunosuppressive effect of mesenchymal stem cell-derived exosomes on a concanavalin A-induced liver injury model. Inflamm Regen. 2016;36:26.
64. Yan Y, Jiang W, Tan Y, Zou S, Zhang H, Mao F, et al. hucMSC exosome-derived GPX1 is required for the recovery of hepatic oxidant injury. Mol Ther. 2017;25:465–79.
65. Haga H, Yan IK, Takahashi K, Matsuda A, Patel T. Extracellular vesicles from bone marrow-derived mesenchymal stem cells improve survival from lethal hepatic failure in mice. Stem Cells Transl Med. 2017;6:1262–72.
66. Bruno S, Herrera Sanchez MB, Pasquino C, Tapparo M, Cedrino M, Tetta C, et al. Huma liver-derived stem cells improve fibrosis and inflammation associated with nonalcoholic steatohepatitis. Stem Cells Int. 2019;2019:6351091.
67. Bruno S, Pasquino C, Herrera Sanchez MB, Tapparo M, Fagiolini F, Grange C, et al. HLSC-derived extracellular vesicles attenuate liver fibrosis and inflammation in a murine model of non-alcoholic steatohepatitis. Mol Ther. 2020;28:479–89.
68. Sun Y, Shi H, Yin S, Ji C, Zhang X, Zhang B, et al. Human mesenchymal stem cell derived exosomes alleviate type 2 diabetes mellitus by reversing peripheral insulin resistance and relieving beta-cell destruction. ACS Nano. 2018;12:7613–28.
69. Srijaya TC, Ramasamy TS, Kasim NHA. Advancing stem cell therapy from bench to bedside: lessons from drug therapies. J Transl Med. 2014;12:243.
70. Povero D, Pinatel EM, Leszczynska A, Goyal NP, Nishio T, Kim J, et al. Human induced pluripotent stem cell-derived extracellular vesicles reduce hepatic stellate cell activation and liver fibrosis. JCI Insight. 2019;5:e125652.

Chapter 9
Current Applications of Endothelial Progenitor Cells–Derived Extracellular Vesicles: From Vascular Repair to Sepsis

Marta Manco and Vincenzo Salemme

1 Introduction

In recent decades, stem cell therapy has emerged as a promising therapeutic approach in several diseases, including type 1 diabetes, heart diseases, Parkinson's disease. Nevertheless, it is noteworthy that the clinical applicability of stem cell therapy is so far hampered by some limitations, including the lack of a standardized method as well as immunogenicity, oncogenic potential, and maldifferentiation of the stem cells once transplanted [1–5]. Indeed, in the last years, some concerns have been raised over its safety. Several studies pointed out that stem cells–mediated tissue repair and regeneration mostly rely on the release of extracellular vesicles (EVs). This finding revolutionized the field of regenerative medicine, thus opening new avenues.

As discussed in the previous chapters, recent mounting evidence demonstrated that EVs function as powerful mediators of intercellular communication, rather than mere cellular fragments or debris. Consistently, EVs carry various proteins and nucleic acids with significant biological roles. As a consequence, the latest developments in biomedical research have laid the foundation for an EV-based cell-free therapy, thus overcoming the stem cell therapy-related limitations. However, there are still some unsolved mysteries and technical hurdles that challenge the field of EV research and their imminent use into the routine clinical practice. Nearly all nucleated cells can release EVs that deliver biomolecules to the recipient cells. Among them, endothelial progenitor cells–derived EVs (EPCs-EVs) exhibited beneficial therapeutic effect in multiple preclinical studies (e.g., cardiovascular

M. Manco (✉) · V. Salemme
Department of Molecular Biotechnology and Health Sciences, Molecular Biotechnology Center, University of Turin, Turin, Italy
e-mail: marta.manco@kuleuven.be; vincenzo.salemme@unito.it

P. V. Pham, S. Fagoonee (eds.), *Stem Cell-Derived Exosomes and Microvesicles*, Stem Cells in Clinical Applications, https://doi.org/10.1007/978-3-031-90698-5_9

diseases, kidney and lung disorders) [6, 7]. This chapter highlights recent advances in unveiling EPCs-EVs (favorable) biological effects, such as anti-inflammatory and angiogenic properties and tissue repair in several diseases, and it will comprehensively discusses the progress in their translation to the clinic.

2 Identification and Definition of Endothelial Progenitor Cells

A pivotal study by Ashara et al. in 1997 showed that endothelial progenitor cells (EPCs) can promote angiogenesis in the context of ischemia [8], providing also key mechanistic insights into the angiogenic process. Notably, 5 years after this first experimental observation, a clinical study aimed at enhancing myocardial regeneration and neovascularization in patients with myocardial infarction was initiated [9]. Overall, both preclinical and clinical studies raised hopes for a potential cure for life-threatening ischemic diseases. Hence, following their discovery, EPC biology was further investigated over the years. Multiple studies demonstrated that EPCs are mononuclear cells that originate from bone marrow and, in the context of pathological conditions, migrate to the damaged sites, and directly contribute to vascularization [10–12]. However, several groups are concerned about EPC identity and function based on other conflicting data [13, 14]. Indeed, over the years, the different isolation and culture methodologies, as well as misleading definitions, have led to confusion in EPC biology, thereby dampening their clinical translation. To this purpose, a lot of effort has been made to standardize EPC nomenclature and to define their cellular phenotype and function [15]. In 2017, a consensus statement shed light on the existence of two major (sub)types of EPCs, which differentiate one from the other for both function (i.e., their respective contribution in supporting angiogenesis) and immunophenotype, as well as for the sequence of appearance and growth capacity in culture (Table 9.1). Therefore, EPCs can be regarded as a highly heterogeneous cellular population. Of note, such heterogeneity explains some conflicting published studies, showing [16] or not [13, 14] their incorporation into the host vasculature.

The major features of "early EPCs" or myeloid angiogenic cells (MACs) are their positivity to hematopoietic and monocyte markers (i.e., CD45 and CD14, respectively, thus suggesting that they are most likely derived from hematopoietic stem cells), limited proliferative capacity, as well as the capability to release angiocrine factors (e.g., VEGF and IL-8) and EVs, thus supporting angiogenesis via paracrine mechanisms [17, 18]. Overall, early EPCs, also known in literature as circulating angiogenic cells (CACs) [19] or "hematopoietic EPCs" [20], exhibit a low vasculogenic potential [21, 22]. On the other hand, "late EPCs" or endothelial colony-forming cells (ECFCs) have a similar phenotype to adult endothelial cells (ECs) and a distinctively strong growth capacity, and they directly contribute to neovascularization in ischemic tissues, by differentiating into mature ECs [17].

Table 9.1 Comparison of early and late EPCs' features

	Early EPCs	Late EPCs
Alias	Hematopoietic EPCs, CACs	Nonhematopoietic EPCs, OECs
Morphology	Spindle-shaped	Cobblestone-like
Appearance in culture	3–5 days	2–4 weeks
Proliferative capacity	High	Low
Surface markers expressed	CD34, CD133, VEGFR2, CD45, CD31, vWE, CD14, CD11c, CD163	VEGFR2, vWE, CD36, CD31, Tie-2, VE-cadherin
AcLDL uptake and lectin binding	Yes	Yes
Migration	Yes	Yes
Tube formation	No	Yes
Cytokine secretion	High	Low
Incorporation into new vessels	No	Yes

Abbreviations: *EPCs* endothelial progenitor cells, *CACs* circulating angiogenic cells, *OECs* outgrowth endothelial cells, *acLDL* acetylated low-density lipoprotein

Hence, late EPCs exhibit a high vasculogenic and angiogenic potential. Furthermore, unlike early EPCs, late EPCs lack the expression of hematopoietic and monocyte markers [23]. Therefore, they are also known as "nonhematopoietic EPCs" [24] or outgrowth endothelial cells (OECs) [25]. Nevertheless, also ECFCs are capable of releasing EVs, crucial for tissue restoration [26–29]. Altogether, early EPCs resemble Asahara's cells [8], whereas late EPCs are similar to those reported in Lin's [30] and Reyes' [31] studies. Although there has been significant debate regarding EPC identity, their role in neovascularization is now clear and undeniable. Hence, both EPCs contribute to the formation of new blood vessels via different mechanisms of action by (1) releasing angiocrine factors, thus instructing neighboring cells and (2) directly differentiating into mature ECs (Fig. 9.1).

3 Endothelial Progenitor Cells Release (Biologically Active) Extracellular Vesicles

EPCs release a large array of paracrine factors (e.g., growth factors, cytokines, chemokines, bioactive lipids) that affect cell behavior in injured tissues, thus ultimately supporting tissue repair and regeneration. These "EPCs-derived factors" have been recently classified into five groups, according to their function [20]. The first group contains neuropilins, semaphorins, VEGFR1, VEGFR2, and VEGFR3 implicated in vasculogenesis and angiogenesis. The second group includes multiple factors (e.g., TNFα, TNFR2/P75, TNFR1/P55, and TRAIL) involved in endothelial-immune cell interaction, proliferation, apoptosis, survival, migration, and immune modulation and immunogenicity. Factors promoting survival, proliferation, migration, and differentiation of vascular stem/progenitor cells (e.g., PDGF, VEGF, TGFβ, FGF2,

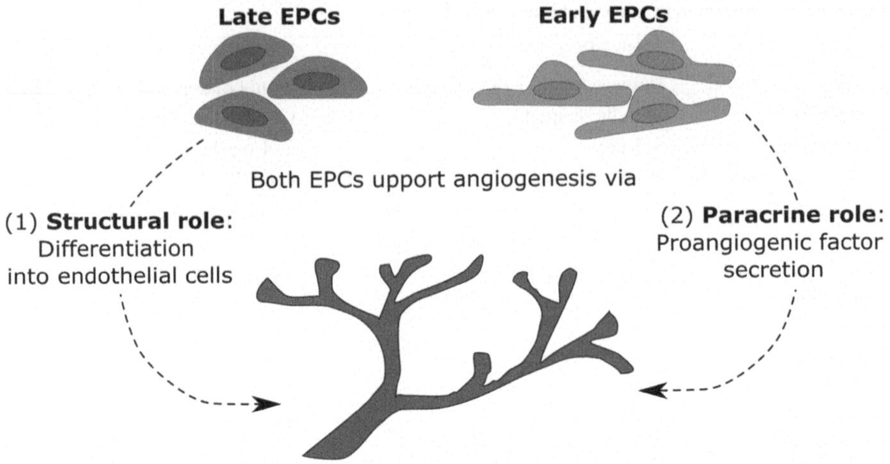

Fig. 9.1 EPCs' role in supporting neovascularization. Early and late EPCs contribute to angiogenesis in a structural and paracrine manner. Early EPCs instruct neighboring cells by releasing angiocrine factor, while late EPCs directly differentiate into mature ECs. *EPCs* endothelial progenitor cells, *ECs* endothelial cells

IGF-1, and EGF) are included in the third group. The fourth group contains non-coding RNA with regulatory activities. The fifth, and last, group includes factors (e.g., uPA, uPAR, uPARAP, tPA, PECAM-1, ICAM-1) implicated in the internalization of ligands, invasiveness, and motility. It is noteworthy to point out that EPCs, like other cell types, can secrete such biomolecules through the release of EVs.

Over recent years, numerous studies have highlighted the therapeutic role of EVs. In this scenario, EVs secreted by stem/progenitor cells can mimic the biological functions of their donor cells, thus promoting tissue repair and regeneration [32]. Likewise, it has been shown that EPCs-derived EVs exert angiogenic effects, by activating an angiogenic program in "resident" ECs via horizontal transfer of biomolecules [33, 34]. Of note, they exhibited a superior therapeutic effect over their ancestors (namely EPCs) on various ischemic diseases [27, 35–37]. These interesting and promising findings paved the way for a valuable cell-free method for regenerative medicine and disease treatment, thus overcoming limitations related to the use of stem/progenitor cells in the clinical practice and, most importantly, preserving the same efficacy and biological output. There are several benefits that distinguish EVs from cell-based therapy in the context of regenerative medicine. For instance, EVs can be exploited as carriers of drug and/or favorable molecules (mainly miRNAs) that promote angiogenesis, fibrosis, and cell proliferation. Moreover, they offer the advantage of being readily available for "off-the-shelf" use and for repetitive transplantation. Importantly, EVs are less immunogenic than their parental cells, making allogenic transplantation feasible. Being produced in large quantities, carrying a similar transcriptome profile of their parental cells, and having a long shelf life, EPCs-derived EVs are considered nowadays an attractive cell-free therapeutic option for either acute or chronic diseases. Subsequently, due to their great potential in clinical translation, several studies provided insights into EVs biogenesis, uptake, and mechanism of action, as well as their long-term preservation and storage. However, this chapter mainly focuses on the functional role of EPCs-derived EVs and their beneficial effects in various animal models for disease and their potential translation to the clinic (Tables 9.2a and 9.2b).

4 EPCs-Derived EVs in Vascular Repair

Transplanted EPCs can migrate to the site of vascular injury site and induce neovascularization, by both releasing EVs and differentiating into mature ECs. Researchers found that EPCs-derived EVs, similar to their ancestor cells, trigger angiogenesis and vascular repair. Deregibus et al., in a pivotal study in 2007, showed that EPCs-derived EVs, expressing on their surface α4 and β1 integrin, can target ECs and transfer their content. In particular, by transferring functional mRNA, EPCs-derived EVs are capable of activating PI3K/AKT/eNOS pathway and, ultimately, an angiogenic program in mature quiescent ECs [34]. Subsequently, several studies further confirm the capability of EPCs-derived EVs to promote re-endothelialization of injured blood vessels.

In-stent restenosis is defined as the reduction in vessel lumen diameter, due to neointima formation, after percutaneous transluminal angioplasty and stenting treatments. This procedure (namely, percutaneous coronary intervention) can lead to ECs damage and vessel injury, thus triggering abnormal proliferation of smooth muscle cells (SMCs), and consequently, neointima hyperplasia. Li et al. determined the therapeutic effects of EPCs-derived EVs in a rat model of balloon-induced vascular injury. Such model mimics the vascular damage and pathological process of restenosis following angioplasty and stenting. The authors showed that EPCs-derived EVs enhance both ECs proliferation and migration, as well as induce the expression of angiogenesis-related genes, thereby promoting vascular repair of the injured carotid artery [38]. Few years later, another study, employing EVs released by human fetal aorta-derived EPCs, strengthened such observation [39]. In addition, the authors showed that EPCs-derived EVs are also capable of reducing SMCs proliferation in vivo [39]. Altogether, EPCs-derived EVs can inhibit neointimal hyperplasia mostly by promoting re-endothelialization, but also by dampening the proliferation of SMCs.

Ischemia-reperfusion injury is accompanied by endothelial hypoxia (H) and re-oxygenation (R) that trigger oxidative stress, thus ultimately leading to vascular dysfunction and endothelial damage. Therefore, a key strategy in the treatment of ischemic stroke consists in protecting ECs from H/R-induced injury. Ma et al. showed that EPCs-derived EVs are capable of reducing H/R-induced apoptosis, reactive oxygen species (ROS) production, and angiogenic dysfunction [59]. Mechanistically, the authors found that normal ECs express higher levels of miR-210, compared to injured ECs by H/R. Hence, miR-210 in EPCs-derived EVs plays a crucial role in promoting vascularization [59]. Likewise, Wang et al.

Table 9.2a Preclinical studies employing EPCs-derived EVs in animal models

Disease	Experimental model	EPCs source	EV delivery	Proposed mechanism/effects of EPCs-derived EVs	References
In-stent restenosis	Balloon-induced vascular injury	Human umbilical cord blood	Intravenously	Increased ECs proliferation, migration, angiogenesis-related molecules	[38]
In-stent restenosis	Balloon-induced vascular injury	Human early fetal aorta	Intravenously	Reendothelialization, reduced neointima hyperplasia, SMCs proliferation	[39]
Traumatic brain injury	Injury inflicted using cortical impact device	Human umbilical cord blood	Intravenously	Preserved BBB integrity: Reduced permeability and edema, as well as preserved tight junctions	[40]
Atherosclerosis	Hypertensive-hyperlipidemic diet	Peripheral blood	Retro-orbital sinus	Reduced atherosclerosis development via miR-10a, miR-21, miR-126 and miR-146a transfer to circulating late EPCs	[41]
Chronic heart failure after infarction	Occlusion of the left anterior descending coronary artery	Bone marrow	Border zone of the infarcted area (encapsulated in gel)	Myocardial remodeling, improved myocardial hemodynamics and angiogenesis	[42]
Cardiac fibrosis	Occlusion of the left anterior descending coronary artery	Peripheral blood	Intramyocardially	Reduced myocardial fibrosis, and enhanced MendoT of cardiac fibroblast via miR-218-5p and miR-363-3p transfer	[43]
Acute kidney injury	Ischemia-reperfusion	Peripheral blood	Intravenously	Increased proliferation and reduced proliferation of tubular cells, as well as reduced capillary rarefaction	[26]
Acute kidney injury	Ischemia-reperfusion	Human umbilical cord	Jugular vein	Reduced ECs apoptosis and ischemic kidney injury via transfer of miR-486-5p targeting PTEN	[29]
Acute kidney injury	Ischemia-reperfusion	Human umbilical cord	Jugular vein	Attenuated ischemic kidney injury via inhibition of ECs apoptosis	[44]
Glomerulonephritis	Anti-Thy1.1 glomerulonephritis	Peripheral blood	Femoral vein	Inhibited antibody- and complement-mediated injury of mesangial cells	[45]

Renal fibrosis	Unilateral ureteral obstruction	Bone marrow	Tail vein	Attenuated renal injury and fibrosis via inhibition of pericyte-myofibroblast transition	[46]
Diabetes	B-cell islet transplantation into SCID mice	Peripheral blood	Subcutaneously	Revascularization of exogenous β-cell islet and increased production/secretion of insulin	[47]
Diabetes	Streptozotocin	Human umbilical cord	Subcutaneously at the wound site	Promotion of cutaneous wound healing (complication of diabetes) and angiogenic program in ECs	[48, 49]

Abbreviations: *EVs* extracellular vesicles, *EPCs* endothelial progenitor cells, *ECs* endothelial cells, *SCID* severe combined immunodeficiency, *BBB* brain blood barrier

Table 9.2b Preclinical studies employing EPCs-derived EVs in animal models

Disease	Experimental model	EPCs source	EV delivery	Proposed mechanism/effects of EPCs-derived EVs	References
Diabetes	Streptozotocin	Bone marrow	Local application on skin wounds	Promotion of skin wound healing, as well as increased angiogenesis-related molecules and proliferation marker Ki-67 (presumably via the transfer of miR-221-3p)	[50]
Diabetes	Db/db mice fed with a high-fat diet	Bone marrow	Intravenously	Reduced atherosclerotic plaques (complication of diabetes) and inflammatory factors, as well as ameliorated endothelium-dependent contractile dysfunction	[51]
Acute lung injury	Lipopolysaccharide	Bone marrow	Intravenously	Restored pulmonary integrity, and improved endothelial function via transfer to ECs of miR-126 targeting SPRED1	[52]
Acute lung injury	Lipopolysaccharide	Human umbilical cord	Intratracheal	Reduced lung injury, permeability, and inflammation via transfer to small airway epithelial cells of miR-126-3p and miR-126-5p targeting PIK3R2, HMGB1 and VEGFα, as well as augmenting tight junction proteins	[53]
Osteoporosis	High-dose dexamethasone	Bone marrow	Tail vein	Prevented osteoporosis by suppressing the ferroptotic pathway in osteoblast	[54]

(continued)

Table 9.2b (continued)

Disease	Experimental model	EPCs source	EV delivery	Proposed mechanism/effects of EPCs-derived EVs	References
Distraction osteogenesis	Unilateral tibial DO	Bone marrow	Local injection into the distraction gaps	Accelerated bone regeneration, and increased angiogenesis via transfer to ECs of miR-126 targeting SPRED1	[55]
Bone fracture	Femur fracture	Mouse umbilical cord	Intravenously	Increased neovascularization at the fracture site and enhanced fracture healing, due to enhanced recruitment and differentiation of osteoclast precursors via transfer of LncRNA-MALAT1-targeting miR-124	[56]
Sepsis	Cecal ligation and puncture	Human umbilical cord	Intravenously	Reduced organ dysfunction, as well as reduced vascular leakage via transfer to ECs of miR-126-3p and miR-126-5p targeting HMGB1 and VCAM1	[57]
Sepsis	Cecal ligation puncture	Human umbilical cord	Intravenously	Reduced sepsis-induced tubular epithelial cells injury via transfer of miR-93-5p targeting KDM6B/TNF-α axis, as well as attenuated multiple organ injury, vascular leakage, and inflammation	[58]

Abbreviations: *EVs* extracellular vesicles, *EPCs* endothelial progenitor cells, *DO* distraction osteogenesis, *ECs* endothelial cells highlighted the protective effects of EPCs-derived EVs in reducing H/R-induced apoptosis and ROS production [60]. The authors elucidated the underlying molecular mechanisms, thus pointing out a pivotal role of miR-126, enriched in EPCs-derived EVs, in activating PI3K/eNOS/NO signaling pathway in brain microvascular ECs [60].

Besides enhancing EC proliferation, migration, survival, and angiogenesis, as well as protecting ECs from ROS overproduction, EPCs-derived EVs can play an important role in repairing the cell-cell connection between damaged ECs. For instance, traumatic brain injury (TBI) causes the disruption of blood-brain barrier (BBB). Gao et al. showed that EPCs-derived EVs, by increasing the expression of tight junction proteins, such as zonula occludens (ZO)-1 and occluding, can ultimately preserve BBB integrity in TBI mouse models [40]. As a consequence, the treatment with EVs alleviated TBI-induced brain edema [40]. Nevertheless, in vitro studies showed EPCs-derived EVs enhance viability, proliferation, migration, and tube formation in brain microvascular ECs. This suggests that EPCs-derived EVs could preserve BBB integrity and functionality through different mechanisms.

5 EPCs-Derived EVs in Cardiovascular Diseases

Cardiovascular diseases (CVDs) account for around 17.9 million deaths each year, thereby being the leading cause of morbidity and mortality worldwide. CVDs, a group of disorders affecting heart or blood vessels, include atherosclerosis, thrombosis, and cardiac fibrosis. In this context, EPCs-derived EVs represent nowadays an attractive and promising therapeutic tool for CVDs and related major complications.

Atherosclerosis is the formation of the so-called plaques inside the lining of the artery wall. These plaques are the consequence of the progressive buildup of cholesterol, fats, and other substances, such as calcium. Over time, these will narrow the artery, thus ultimately blocking blood flow. In 2020, Alexandru et al. demonstrated the capability of EPCs-derived EVs in dampening atherosclerotic vascular disease [41]. In particular, the authors isolated EPCs from peripheral blood of healthy hamsters and injected the related vesicles in hypertensive and hyperlipidemic hamsters. They found that EPCs-derived EVs significantly reduce atherosclerosis development by mitigating (1) dyslipidemia and hypertension, (2) circulating cytokine and chemokine levels, and (3) the structural and functional remodeling of arterial and left ventricular walls. In addition, the underlying molecular mechanisms were elucidated. EPCs-derived EVs contain and transfer biologically active miRNAs (i.e., miR-10a, miR-21, miR-126, and miR-146a) to circulating late EPCs, thus protecting them against atherosclerotic injury and mediating vascular repair and regeneration [41].

The biological role and potential therapeutic effects of EPCs-derived EVs have also been studied in other pathological contexts, such as chronic heart failure following myocardial infarction. Consistently, Chen et al. provided the first evidence that the treatment with EPCs-derived EVs improves cardiac function in rats with myocardial infarction [42]. Of note, the authors improved the delivery of EVs by encapsulating them into a shear-thinning gel, which was precisely injected at the border of the infarcted area. In vivo animal experiments demonstrated that EVs enhanced both peri-infarct angiogenesis and myocardial hemodynamics [42]. Another subsequent study further strengthened these findings [61].

Cardiac fibrosis refers to the excessive accumulation of extracellular matrix in the cardiac muscle, thus leading over time to both structural and functional abnormalities. Myocardial fibrosis occurs when activated fibroblasts proliferate and secrete extracellular matrix components. Moreover, in the literature, it has been well recognized that (1) ECs undergo endothelial-to-mesenchymal transition, thus contributing to a subset of cardiac fibroblast, as well as, although to a smaller extent, (2) cardiac fibroblast can adopt an EC fate, thus contributing to neovascularization and repair of the injured heart [62]. A couple of works demonstrated that EPCs-derived EVs in vitro promote mesenchymal-to-endothelial transition (MEndoT) of cardiac fibroblasts, thereby promoting their proliferation and angiogenesis [63, 64]. Notably, miRNAs played a critical role in EVs-mediated MEndoT induction of cardiac fibroblasts [63]. EPCs-derived EVs are particularly enriched of miR-133, which has been shown to increase the expression of endothelial markers (e.g., CD31, vWF, and VE-cadherin) and, on the other hand, to reduce mesenchymal markers (e.g., Vimentin, αSMA, and Collagen I) in cardiac fibroblast [63]. Likewise, it has been demonstrated that (normoxic) EPCs-derived EVs attenuate TGFβ-dependent activation of cardiac fibroblast [65]. Mechanistically, miR-10b-5p, particularly enriched in such EVs, targets the fibrotic genes *SMURF1* and *HDAC4* in the recipient cardiac fibroblast [65]. Another study shed light on the biological role of miRNAs-enriched EVs in enhancing angiogenesis and inhibiting myocardial fibrosis [43]. For instance, Ke et al. demonstrated that EPCs-derived EVs, by transferring miR-218-5p and miR-363-3p to cardiac fibroblast, induce their proliferation and angiogenesis. Consistently, the administration of EPCs-derived EVs in a rat with myocardial infarction inhibited myocardial fibrosis, as well as restored myocardial tissue integrity [43]. Altogether, these results point out that EPCs-derived EVs can inhibit fibroblast activation and favor MEndoT occurring in cardiac fibrosis, thus strongly suggesting their employment in therapeutic protocols. Nevertheless, further in vivo experiments need to be performed to confirm and strengthen these promising in vitro data.

Cardiac hypertrophy-dependent heart failure is the major cause of cardiovascular morbidity and mortality. Among the underlying pathophysiological mechanisms, there are cardiomyocyte hypertrophy, apoptosis, oxidative stress, and inflammation. For instance, the concept that angiotensin (Ang) II-induced oxidative stress and inflammation has gained general acceptance [66, 67]. Gu et al. demonstrated that EPCs-derived EVs can prevent Ang-II-induced apoptosis of cardiomyocytes, hypertrophy (as suggested by their reduced cell size and the expression of the fetal cardiac gene β-MHC, upon EVs treatment), and ROS production [68]. Interestingly, RNA depletion from EVs eliminated their beneficial effects [68]. This strongly suggests that, also in this case, RNA is a key player in mediating the biological role of vesicles.

6 EPCs-Derived EVs in Kidney Diseases

Ischemia-reperfusion is the main cause of acute kidney injury (AKI), with the latter being associated with a high risk of developing chronic kidney diseases (CKDs). It is accompanied by tubular cell necrosis and vascular dysfunction or EC loss. Over the years, growing evidence has pointed out the therapeutic effects of EPCs-derived EVs in alleviating AKI, thereby dampening the development of CKDs. For instance, Cantaluppi et al. determined the therapeutic potential of EPCs-derived EVs in an AKI rat model [26]. Vesicles were mainly localized within tubular cells and peritubular ECs, thus enhancing tubular cell proliferation, reducing apoptosis, and inhibiting capillary refraction. Altogether, serum biochemistry analysis (i.e., creatinine and BUN levels) and histological examination revealed an improvement of kidney injury, from both structural and functional points of view, upon EVs treatment. Consequently, CKD progression, namely leukocyte infiltration, glomerulosclerosis, and tubulointerstitial fibrosis, was counteracted [26]. Mechanistically, the renoprotective effect of EVs relied on an RNA-dependent reprogramming of the resident renal cells. In particular, miR-126 and miR-296, enriched in EPCs-derived EVs, were identified as the key molecules [26]. Of note, such miRNAs were previously found to exert beneficial effects in a murine model of hindlimb ischemia [27]. Subsequently, the therapeutic potential of EPCs-derived EVs in experimental AKI models was further strengthened in other works [29, 44]. For instance, it has been demonstrated that EVs transfer miR-486-5p to resident cells, thereby reducing ischemic kidney injury [29].

Glomerulonephritis refers to the immune-mediated glomerular inflammation, which in turn causes acute kidney injury and/or chronic renal failure, thus hampering glomerular filtration. Unfortunately, no effective treatment is so far available for this disease. However, multiple studies pointed out the beneficial role of EPCs and related EVs in glomerulonephritis animal models, such as the mesangiolytic anti-Thy1.1 glomerulonephritis (GN) model [45, 69, 70]. Among them, a study by Cantaluppi et al. showed that the treatment with EPCs-derived EVs significantly reduced tissue damage, thus ultimately preserving renal function [45]. Mechanistically, the renoprotective effect was attributed to several mRNAs (e.g., encoding Factor H, CD55, and CD59). Consistently, EPCs-derived EVs might play a role in repairing tissue injury by transferring their RNAs and, ultimately, promoting angiogenesis and counteracting complement-mediated injury of mesangial cells.

Renal fibrosis is the excessive accumulation of extracellular matrix and the main pathological feature of CKDs. The beneficial effect of EPCs-derived EVs has been further evaluated in a renal fibrosis model induced by unilateral ureteral obstruction (UUO). In this context, Yang et al. showed that EPCs-derived EVs attenuated renal fibrosis by dampening pericyte-myofibroblast transition without promoting vascular repair [46]. Indeed, it is noteworthy that the impact of EVs on the other cell types, besides ECs, was the most prominent in this specific pathological context. This strongly corroborates the concept that EPCs-derived EVs not only have beneficial effects on ECs, thus triggering vascular repair and angiogenesis, but also on other cell types, which are involved in the pathological processes underlying kidney diseases. Altogether, these studies pointed out the great potential of EPCs-derived EVs in orchestrating renal regeneration upon acute or chronic insults.

7 EPCs-Derived EVs in Diabetes

Type 1 diabetes mellitus results from the immune-mediated destruction of insulin-producing β-cells in the pancreas. Consistently, islet cell transplantation has been considered a valuable strategy to restore β-cell function in such diseases. However, loss of transplanted islets, mostly by ischemia due to vascular dysplasia, is very common. Therefore, a lot of effort has been made to counteract the poor graft vascularization and, consequently, to improve the success rate of islet transplantation. For instance, EPCs-derived EVs, which exhibited protective effects in ischemia-reperfusion injury and cardiovascular diseases, are nowadays considered a valuable therapeutic tool to enhance islet vascularization. Indeed, Cantaluppi et al. transplanted islet endothelial cells and β-cells into severe combined immunodeficiency (SCID) mice and uncovered the effect and the underlying molecular mechanisms of EPCs-derived EVs, by injecting them intravenously [47]. EVs, taken up by transplanted cells, promoted insulin secretion, survival, and revascularization of exogenous islets. Of note, EPCs-derived EVs positively regulate EC behavior, thus promoting their proliferation, migration, and organization in vessel-like structures [47]. In addition, EPCs-derived EVs promote transplanted β-cells-dependent insulin production and secretion. Mechanistically, the authors related these biological effects to (1) the activation of PI3K/AKT/eNOS signaling pathway and to the (2) transfer of miR-126 and miR-296.

Chronic skin damage and atherosclerosis are among the major complications of diabetes and heavily contribute to lower both life quality and expectancy of patients. Therefore, efficient therapeutic strategies are urgently needed. EPCs-derived EVs have been shown to fully meet these clinical needs [48–51]. Concerning diabetic ulcers, EPCs-derived EVs can facilitate skin wound healing via positive modulation of EC behavior. Consistently, Li et al. showed that such EVs can promote angiogenesis in the damaged areas and, consequently, skin wound healing in diabetic rats [48]. Few months later, another study further corroborated these findings and pointed out that EPCs-derived EVs positively modulate EC function by acting on ERK1/2 signaling pathway [49]. Over the years, it has been reported that the biological role of EPCs-derived EVs mostly relies on the horizontal transfer of their miRNAs. For this reason, the role of EVs-enriched miRNAs in healing diabetic skin wounds has been recently explored [50]. Interestingly, the authors found that miRNA-221-3p is one of the most abundant miRNAs in EPCs-derived EVs and that it promotes skin wound healing in diabetic mice [50]. Mechanistically, miRNA-221-3p led to an increase in the expression of angiogenesis-related genes, such as VEGF and CD31, and proliferation marker Ki-67 [50].

Moreover, EPCs-derived EVs might have a therapeutic potential in the context of atherosclerosis by counteracting endothelial dysfunction (preceding atherosclerosis development) [51]. The authors showed in detail that EPCs-derived EVs, by transferring their specific regulatory miRNAs (i.e., miR-21a-5p, miR-222-3p, miR-221-3p, miR-155-5p, and miR-29a-3p), restore endothelial functions in mice with diabetic atherosclerosis. On one side, the treatment with EPCs-derived EVs significantly decreased atherosclerotic plaque formation and production of inflammatory factors and on the other hand improved the endothelium-dependent contractility of thoracic aorta [51]. Altogether, EPCs-derived EVs could provide an essential therapeutic tool to improve clinical outcome of diabetic patients, by (1) restoring β-cell function and (2) attenuating the appearance of the most common complications.

8 EPCs-Derived EVs in Lung Injury

Over the years, multiple studies highlighted the therapeutic properties of EPCs-derived EVs in a large array of pathophysiological contexts, such as lung injury. Both acute lung injury (ALI) and acute respiratory distress syndrome (ARDS), despite being caused by different etiologies, have a similar clinical picture, and consist in the extensive injury of the alveolar epithelial cells and capillary ECs. Consequently, protein-rich edema fluid influx into the air spaces occurs, thus causing acute hypoxic respiratory insufficiency. A vicious circle, consisting of ECs damage, increased pulmonary vascular permeability, pulmonary edema, hypoxemia, and again ECs damage, will be established, thus emphasizing that endothelial repair is crucial to prevent ALI/ARDS propagation. Wu et al. demonstrated that EPCs-derived EVs alleviated LPS-induced lung injury by (1) restoring pulmonary capillary integrity and (2) reducing both pulmonary inflammation and edema formation [52]. The authors provided the related mechanistic insights and concluded that miR-126, particularly enriched in EPCs-derived EVs, downregulates SPRED1 and activates RAF/ERK signaling pathway, thus promoting EC proliferation, migration, and angiogenesis [52]. Indeed, miR-126 knockdown in EPCs prevented the therapeutic effect of EPCs-derived EVs on LPS-treated endothelial cells. Another study further corroborated the beneficial role of EPCs-derived EVs in LPS-induced lung injury mice by delivering such miRNA [53]. However, the authors proposed a different underlying cellular and molecular mechanisms. Besides the effect on ECs, it has been demonstrated that EPCs-derived EVs and miRNA-126 exhibited anti-inflammatory properties, by reducing the alveolar-capillary barrier damage, the cytokine/chemokine levels in the bronchoalveolar lavage fluid and lung neutrophil infiltration. Mechanistically, the highly abundant miR-126-3p and miR-126-5p can target and downregulate PI3KR2, HMGB1 VEGFα, as well as can attenuate the LPS-induced downregulation of tight junction proteins (i.e., occludin, claudin1, and claudin4), thus preserving the alveolar epithelial barrier integrity in the lung and, consequently, mitigating alveolar edema [53]. Altogether, although further studies are needed, these works pointed out the therapeutic potential of EPCs-derived EVs in human pulmonary diseases.

9 EPCs-Derived EVs in Bone Healing and Diseases

Multiple studies pointed out the broad therapeutic advantage of EPCs-derived EVs, thereby extending their application to cure other various diseases, such as osteoporosis. Osteoporosis is a disease that makes bones fragile and more likely to fracture. It develops slowly over several years and is the leading cause of fractures in the elderly. Nevertheless, the extensive chronic use of steroids, which are indicated in many pediatric medical conditions, increased the incidence of osteoporosis in young patients. By harnessing the steroid-induced osteoporosis mouse model, Lu et al. found that EPCs-derived EVs prevented mouse glucocorticoid-induced osteoporosis by negatively regulating ferroptosis and by alleviating oxidative injury in osteoblasts [54].

Distraction osteogenesis (DO) is a surgical technique used to promote bone tissue regeneration by making a cut in the bone and allowing the healing process to gradually fill the created gap. This process will lead to the prolongation and defect corrections of the bone. EPCs are mobilized from bone marrow into peripheral circulation and contribute to new bone formation in fracture healing and DO, mostly by promoting neovascularization [71]. For this reason, the effect of EPCs-derived EVs was evaluated in rat model of tibial DO. The authors found that EVs treatment can stimulate angiogenesis and effectively promote bone regeneration during DO procedure [55]. Indeed, angiogenesis and osteogenesis are closely coupled. This means that EPCs-derived EVs treatment can be used as an adjuvant therapy to shorten the duration of DO.

Bone repair involves osteoclasts-mediated bone resorption, osteoblast-mediated formation of new bone, and neovascularization. Multiple studies highlighted the role of EPCs-derived EVs at multiple levels in bone repair process. For instance, EPCs-derived EVs can promote bone regeneration by affecting bone marrow stromal cells' osteoblastic differentiation and proliferation [72]. Moreover, Chen et al. demonstrated that EPCs-derived EVs stimulate osteoblast proliferation and migration via the miR-126/ERK1/2-Bcl-2 pathway, thus promoting bone regeneration [73]. Cui et al. found that EPCs-derived EVs enhance recruitment and differentiation of osteoclast progenitor cells (i.e., osteoclastogenesis) through lncRNA-MALAT1. The latter directly binds and sequesters miR-214, a negative regulator of progenitor cell differentiation into osteoclasts, thus ultimately interfering with its activity [56]. Altogether, these studies further corroborated the hypothesis that

EPCs-derived EVs could offer a new therapeutic approach in the context of bone regeneration and fracture healing.

10 EPCs-Derived EVs in Neurodegenerative Disorders

Although few studies have been so far performed on EPCs and brain diseases, the encouraging results obtained suggest EPCs-derived EVs as a promising treatment strategy for amyotrophic lateral sclerosis (ALS). ALS is characterized by EC alteration and the destruction of the blood-spinal cord barrier (BSCB). The maintenance of BSCB integrity and functionality could be crucial to counteract motor neuron dysfunction and to alleviate the propagation of ALS. For instance, Garbuzova-Davis et al. provided the first evidence that the transplantation of EPCs into symptomatic G93A SOD1 mice, which resemble the features of human ALS, effectively repairs BSCB, thereby preventing the entry of immune/inflammatory cells and preserving motor neuron survival and functionality [74]. Subsequently, the same group tested a cell-free therapeutic approach for endothelium repair in ALS. More in detail, the authors mimicked the pathological environmental conditions of ALS by exposing mouse brain ECs to ALS mice-derived plasma. Of note, they found that EPCs-derived EVs reduce mouse brain EC damage [75]. However, despite being interesting and attractive results, to determine the translation potential of EPCs-derived EVs in vivo experiments are needed. Of note, EVs, which are able to cross the BBB and BSCB [76], might represent a promising candidate for the treatment of a large array of neurodegenerative disorders.

11 EPCs-Derived EVs in Sepsis

Sepsis is a potentially life-threatening condition characterized by systemic inflammation, primarily triggered by harmful microorganisms, which can lead to multi-organ dysfunction.

Growing evidence highlighted the therapeutic effect of EPCs in animal models of sepsis [77, 78]. Moreover, EPCs-derived EVs exhibit protective effect on microvasculature after sepsis, mainly through the delivery of miR-126-3p and miR-126-5p [57]. Sepsis is characterized by PAMPs- and DAMPs-dependent induction of pro-inflammatory cytokines, which in turn cause vascular injury and increased permeability. Of note, miR-126-3p and miR-126-5p, highly abundant in EPCs-derived EVs, suppressed DAMPs-dependent induction of HMGB1 and VCAM1 in microvascular ECs, thus preventing adverse septic complications [57]. Indeed, EVs treatment improves the outcome (e.g., survival, lung and renal vascular leakage, liver, and kidney dysfunction) of septic mice.

Another study further confirmed the therapeutic effect of EPCs-derived EVs in sepsis-induced AKI and shed light on the underlying molecular mechanism [58]. It is not surprising that, even in this context, the biological role of EPCs-derived EVs relies on the delivery of their miRNAs. In particular, EPCs-derived EVs carry miR-93-5p that, by regulating KDM6BH/3K27me3/TNF-α axis, exerts protective effects. Indeed, in vitro experiments proved that miR-93-59 directly binds KDM6B, thus leading to (1) KDM6B silencing, (2) H3K27me3 induction and, ultimately, (3) inhibition of TNF-α. Overall, miR-93-5p can weaken LPS-induced injury of tubular epithelial cells [58]. These data were further corroborated by exploiting the cecal ligation and puncture (CLP) septic mouse model. Consistently, He et al. found that EPCs-derived EVs, carrying miR-93-5p, improve survival and alleviate injury, vascular leakage, inflammation, and apoptosis in multiple organs of septic mice [58].

12 Conclusions: Application of EPCs-Derived EVs in Regenerative Medicine

Regenerative medicine aims at restoring the physiological function of damaged and compromised tissues/organs. In this regard, EVs hold significant promise for achieving such functional tissue regeneration. The therapeutic potential of EPCs-derived EVs has been highlighted in a large array of studies, thus being regarded as one of the most attractive tools in this field nowadays (Fig. 9.2). As previously discussed, EPCs-EVs can promote tissue repair and regeneration in a paracrine manner by regulating migration, proliferation, and differentiation, as well as reprogramming the phenotype of recipient cells. Indeed, EPC-EVs can be internalized into recipient cells and can then release their content, such as proteins, lipids, mRNAs, and miRNAs. Based on this rationale, many efforts have also been made to load EPCs-derived EVs with specific factors (e.g., miR-210, miR-137, miR-126, osteocalcin), thus harnessing EVs as carrier and reinforcing their capability to induce reprograming of damaged recipient cells, and ultimately tissue regeneration [59, 73, 79–81]. This strongly pinpoints the extraordinary therapeutic potential of both naïve and modified/loaded EPCs-derived EVs. However, despite EVs being considered a novel therapeutic tool in several pathological contexts, translation to the clinic has been so far modest. The major limitations that currently challenge the full realization of their therapeutic utility are (1) scalability and (2) standardization of production, characterization, and targeted delivery.

Acknowledgement We thank Dr. Sharmila Fagoonee for the fruitful scientific discussion and for the revision of this work. In this study, several sentences were improved using ChatGPT, a language model created by OpenAI for text generation and linguistic analysis. ChatGPT is a state-of-the-art natural language processing model built on the GPT-3.5 architecture (OpenAI, 2022).

Conflict of Interest The authors declare no conflicts of interest.

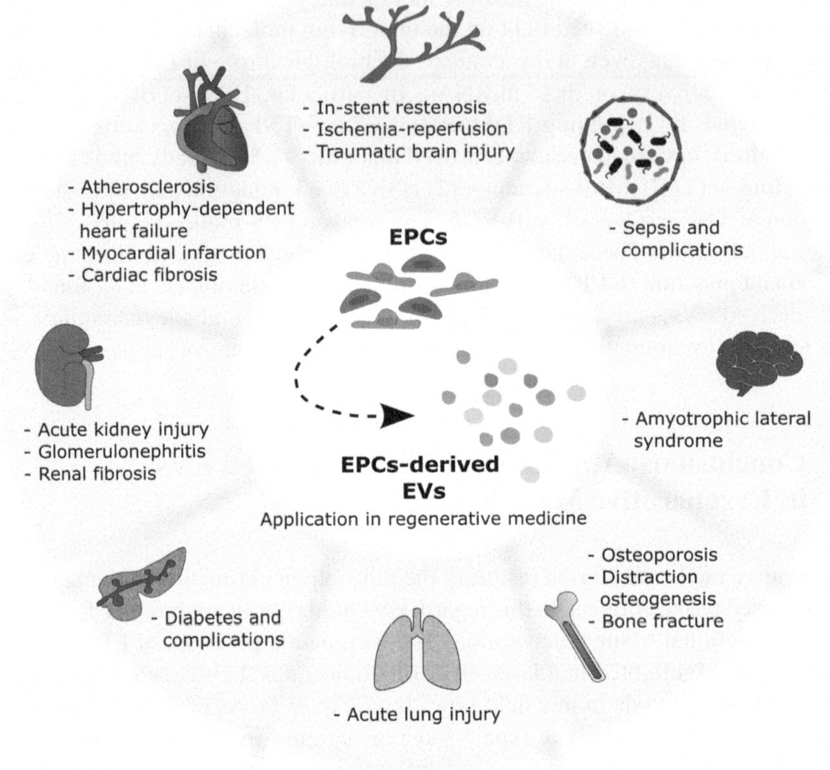

Fig. 9.2 Applications of EPCs-derived EVs in regenerative medicine. Multiple preclinical in vitro and in vivo studies pointed out the therapeutic effect of EPCs-derived EVs in several pathological contexts, spanning from CVDs to brain injury. EVs can nowadays be regarded as biologically active cellular products that, by transferring miRNAs and other functional molecules, play a key role in intercellular communication. Consistently, EVs exhibit regenerative, anti-inflammatory, antioxidant, and antifibrotic features, thus providing us with a novel and promising therapeutic possibility. *EPCs* endothelial progenitor cells, *EVs* extracellular vesicles, *CVDs* cardiovascular diseases

References

1. Amariglio N, Hirshberg A, Scheithauer BW, Cohen Y, Loewenthal R, Trakhtenbrot L, et al. Donor-derived brain tumor following neural stem cell transplantation in an ataxia telangiectasia patient. PLoS Med. 2009;6(2):e1000029. https://doi.org/10.1371/journal.pmed.1000029.
2. Werbowetski-Ogilvie TE, Bossé M, Stewart M, Schnerch A, Ramos-Mejia V, Rouleau A, et al. Characterization of human embryonic stem cells with features of neoplastic progression. Nat Biotechnol. 2009;27(1):91–7. https://doi.org/10.1038/nbt.1516.
3. Aguilar S, Nye E, Chan J, Loebinger M, Spencer-Dene B, Fisk N, et al. Murine but not human mesenchymal stem cells generate osteosarcoma-like lesions in the lung. Stem Cells. 2007;25(6):1586–94. https://doi.org/10.1634/stemcells.2006-0762.

4. Breitbach M, Bostani T, Roell W, Xia Y, Dewald O, Nygren JM, et al. Potential risks of bone marrow cell transplantation into infarcted hearts. Blood. 2007;110(4):1362–9. https://doi.org/10.1182/blood-2006-12-063412.
5. Kunter U, Rong S, Boor P, Eitner F, Müller-Newen G, Djuric Z, et al. Mesenchymal stem cells prevent progressive experimental renal failure but maldifferentiate into glomerular adipocytes. J Am Soc Nephrol. 2007;18(6):1754–64. https://doi.org/10.1681/ASN.2007010044.
6. Xing Z, Zhao C, Liu H, Fan Y. Endothelial progenitor cell-derived extracellular vesicles: a novel candidate for regenerative medicine and disease treatment. Adv Healthc Mater. 2020;9(12):e2000255. https://doi.org/10.1002/adhm.202000255.
7. Terriaca S, Fiorelli E, Scioli MG, Fabbri G, Storti G, Cervelli V, et al. Endothelial progenitor cell-derived extracellular vesicles: potential therapeutic application in tissue repair and regeneration. Int J Mol Sci. 2021;22(12) https://doi.org/10.3390/ijms22126375.
8. Asahara T, Murohara T, Sullivan A, Silver M, van der Zee R, Li T, et al. Isolation of putative progenitor endothelial cells for angiogenesis. Science. 1997;275(5302):964–7. https://doi.org/10.1126/science.275.5302.964.
9. Strauer BE, Brehm M, Zeus T, Köstering M, Hernandez A, Sorg RV, et al. Repair of infarcted myocardium by autologous intracoronary mononuclear bone marrow cell transplantation in humans. Circulation. 2002;106(15):1913–8. https://doi.org/10.1161/01.cir.0000034046.87607.1c.
10. Asahara T, Masuda H, Takahashi T, Kalka C, Pastore C, Silver M, et al. Bone marrow origin of endothelial progenitor cells responsible for postnatal vasculogenesis in physiological and pathological neovascularization. Circ Res. 1999;85(3):221–8. https://doi.org/10.1161/01.res.85.3.221.
11. Bailey AS, Jiang S, Afentoulis M, Baumann CI, Schroeder DA, Olson SB, et al. Transplanted adult hematopoietic stems cells differentiate into functional endothelial cells. Blood. 2004;103(1):13–9. https://doi.org/10.1182/blood-2003-05-1684.
12. Peters BA, Diaz LA, Polyak K, Meszler L, Romans K, Guinan EC, et al. Contribution of bone marrow-derived endothelial cells to human tumor vasculature. Nat Med. 2005;11(3):261–2. https://doi.org/10.1038/nm1200.
13. Göthert JR, Gustin SE, van Eekelen JA, Schmidt U, Hall MA, Jane SM, et al. Genetically tagging endothelial cells in vivo: bone marrow-derived cells do not contribute to tumor endothelium. Blood. 2004;104(6):1769–77. https://doi.org/10.1182/blood-2003-11-3952.
14. Purhonen S, Palm J, Rossi D, Kaskenpää N, Rajantie I, Ylä-Herttuala S, et al. Bone marrow-derived circulating endothelial precursors do not contribute to vascular endothelium and are not needed for tumor growth. Proc Natl Acad Sci USA. 2008;105(18):6620–5. https://doi.org/10.1073/pnas.0710516105.
15. Medina RJ, Barber CL, Sabatier F, Dignat-George F, Melero-Martin JM, Khosrotehrani K, et al. Endothelial progenitors: a consensus statement on nomenclature. Stem Cells Transl Med. 2017;6(5):1316–20. https://doi.org/10.1002/sctm.16-0360.
16. Yoder MC, Mead LE, Prater D, Krier TR, Mroueh KN, Li F, et al. Redefining endothelial progenitor cells via clonal analysis and hematopoietic stem/progenitor cell principals. Blood. 2007;109(5):1801–9. https://doi.org/10.1182/blood-2006-08-043471.
17. Hur J, Yoon CH, Kim HS, Choi JH, Kang HJ, Hwang KK, et al. Characterization of two types of endothelial progenitor cells and their different contributions to neovasculogenesis. Arterioscler Thromb Vasc Biol. 2004;24(2):288–93. https://doi.org/10.1161/01.ATV.0000114236.77009.06.
18. O'Neill CL, Guduric-Fuchs J, Chambers SE, O'Doherty M, Bottazzi B, Stitt AW, et al. Endothelial cell-derived pentraxin 3 limits the vasoreparative therapeutic potential of circulating angiogenic cells. Cardiovasc Res. 2016;112(3):677–88. https://doi.org/10.1093/cvr/cvw209.
19. Rehman J, Li J, Orschell CM, March KL. Peripheral blood "endothelial progenitor cells" are derived from monocyte/macrophages and secrete angiogenic growth factors. Circulation. 2003;107(8):1164–9. https://doi.org/10.1161/01.cir.0000058702.69484.a0.

20. Abdelgawad ME, Desterke C, Uzan G, Naserian S. Single-cell transcriptomic profiling and characterization of endothelial progenitor cells: new approach for finding novel markers. Stem Cell Res Ther. 2021;12(1):145. https://doi.org/10.1186/s13287-021-02185-0.
21. Mukai N, Akahori T, Komaki M, Li Q, Kanayasu-Toyoda T, Ishii-Watabe A, et al. A comparison of the tube forming potentials of early and late endothelial progenitor cells. Exp Cell Res. 2008;314(3):430–40. https://doi.org/10.1016/j.yexcr.2007.11.016.
22. Sieveking DP, Buckle A, Celermajer DS, Ng MK. Strikingly different angiogenic properties of endothelial progenitor cell subpopulations: insights from a novel human angiogenesis assay. J Am Coll Cardiol. 2008;51(6):660–8. https://doi.org/10.1016/j.jacc.2007.09.059.
23. Bouvard C, Gafsou B, Dizier B, Galy-Fauroux I, Lokajczyk A, Boisson-Vidal C, et al. alpha6-integrin subunit plays a major role in the proangiogenic properties of endothelial progenitor cells. Arterioscler Thromb Vasc Biol. 2010;30(8):1569–75. https://doi.org/10.1161/ATVBAHA.110.209163.
24. Chopra H, Hung MK, Kwong DL, Zhang CF, Pow EHN. Insights into endothelial progenitor cells: origin, classification, potentials, and prospects. Stem Cells Int. 2018;2018:9847015. https://doi.org/10.1155/2018/9847015.
25. Gulati R, Jevremovic D, Peterson TE, Chatterjee S, Shah V, Vile RG, et al. Diverse origin and function of cells with endothelial phenotype obtained from adult human blood. Circ Res. 2003;93(11):1023–5. https://doi.org/10.1161/01.RES.0000105569.77539.21.
26. Cantaluppi V, Gatti S, Medica D, Figliolini F, Bruno S, Deregibus MC, et al. Microvesicles derived from endothelial progenitor cells protect the kidney from ischemia-reperfusion injury by microRNA-dependent reprogramming of resident renal cells. Kidney Int. 2012;82(4):412–27. https://doi.org/10.1038/ki.2012.105.
27. Ranghino A, Cantaluppi V, Grange C, Vitillo L, Fop F, Biancone L, et al. Endothelial progenitor cell-derived microvesicles improve neovascularization in a murine model of hindlimb ischemia. Int J Immunopathol Pharmacol. 2012;25(1):75–85. https://doi.org/10.1177/039463201202500110.
28. Viñas JL, Spence M, Gutsol A, Knoll W, Burger D, Zimpelmann J, et al. Receptor-ligand interaction mediates targeting of endothelial Colony forming cell-derived exosomes to the kidney after ischemic injury. Sci Rep. 2018;8(1):16320. https://doi.org/10.1038/s41598-018-34557-7.
29. Viñas JL, Burger D, Zimpelmann J, Haneef R, Knoll W, Campbell P, et al. Transfer of microRNA-486-5p from human endothelial colony forming cell-derived exosomes reduces ischemic kidney injury. Kidney Int. 2016;90(6):1238–50. https://doi.org/10.1016/j.kint.2016.07.015.
30. Lin Y, Weisdorf DJ, Solovey A, Hebbel RP. Origins of circulating endothelial cells and endothelial outgrowth from blood. J Clin Invest. 2000;105(1):71–7. https://doi.org/10.1172/JCI8071.
31. Reyes M, Dudek A, Jahagirdar B, Koodie L, Marker PH, Verfaillie CM. Origin of endothelial progenitors in human postnatal bone marrow. J Clin Invest. 2002;109(3):337–46. https://doi.org/10.1172/JCI14327.
32. De Jong OG, Van Balkom BW, Schiffelers RM, Bouten CV, Verhaar MC. Extracellular vesicles: potential roles in regenerative medicine. Front Immunol. 2014;5:608. https://doi.org/10.3389/fimmu.2014.00608.
33. Johnson TK, Zhao L, Zhu D, Wang Y, Xiao Y, Oguljahan B, et al. Exosomes derived from induced vascular progenitor cells promote angiogenesis in vitro and in an in vivo rat hindlimb ischemia model. Am J Physiol Heart Circ Physiol. 2019;317(4):H765–H76. https://doi.org/10.1152/ajpheart.00247.2019.
34. Deregibus MC, Cantaluppi V, Calogero R, Lo Iacono M, Tetta C, Biancone L, et al. Endothelial progenitor cell derived microvesicles activate an angiogenic program in endothelial cells by a horizontal transfer of mRNA. Blood. 2007;110(7):2440–8. https://doi.org/10.1182/blood-2007-03-078709.

35. Sahoo S, Klychko E, Thorne T, Misener S, Schultz KM, Millay M, et al. Exosomes from human CD34(+) stem cells mediate their proangiogenic paracrine activity. Circ Res. 2011;109(7):724–8. https://doi.org/10.1161/CIRCRESAHA.111.253286.
36. Bitzer M, Ben-Dov IZ, Thum T. Microparticles and microRNAs of endothelial progenitor cells ameliorate acute kidney injury. Kidney Int. 2012;82(4):375–7. https://doi.org/10.1038/ki.2012.152.
37. Mathiyalagan P, Liang Y, Kim D, Misener S, Thorne T, Kamide CE, et al. Angiogenic mechanisms of human CD34. Circ Res. 2017;120(9):1466–76. https://doi.org/10.1161/CIRCRESAHA.116.310557.
38. Li X, Chen C, Wei L, Li Q, Niu X, Xu Y, et al. Exosomes derived from endothelial progenitor cells attenuate vascular repair and accelerate reendothelialization by enhancing endothelial function. Cytotherapy. 2016;18(2):253–62. https://doi.org/10.1016/j.jcyt.2015.11.009.
39. Kong J, Wang F, Zhang J, Cui Y, Pan L, Zhang W, et al. Exosomes of endothelial progenitor cells inhibit Neointima formation after carotid artery injury. J Surg Res. 2018;232:398–407. https://doi.org/10.1016/j.jss.2018.06.066.
40. Gao W, Li F, Liu L, Xu X, Zhang B, Wu Y, et al. Endothelial colony-forming cell-derived exosomes restore blood-brain barrier continuity in mice subjected to traumatic brain injury. Exp Neurol. 2018;307:99–108. https://doi.org/10.1016/j.expneurol.2018.06.001.
41. Alexandru N, Andrei E, Safciuc F, Dragan E, Balahura AM, Badila E, et al. Intravenous administration of allogenic cell-derived microvesicles of healthy origins defend against atherosclerotic cardiovascular disease development by a direct action on endothelial progenitor cells. Cells. 2020;9(2) https://doi.org/10.3390/cells9020423.
42. Chen CW, Wang LL, Zaman S, Gordon J, Arisi MF, Venkataraman CM, et al. Sustained release of endothelial progenitor cell-derived extracellular vesicles from shear-thinning hydrogels improves angiogenesis and promotes function after myocardial infarction. Cardiovasc Res. 2018;114(7):1029–40. https://doi.org/10.1093/cvr/cvy067.
43. Ke X, Yang R, Wu F, Wang X, Liang J, Hu X, et al. Exosomal miR-218-5p/miR-363-3p from endothelial progenitor cells ameliorate myocardial infarction by targeting the p53/JMY signaling pathway. Oxidative Med Cell Longev. 2021;2021:5529430. https://doi.org/10.1155/2021/5529430.
44. Burger D, Viñas JL, Akbari S, Dehak H, Knoll W, Gutsol A, et al. Human endothelial colony-forming cells protect against acute kidney injury: role of exosomes. Am J Pathol. 2015;185(8):2309–23. https://doi.org/10.1016/j.ajpath.2015.04.010.
45. Cantaluppi V, Medica D, Mannari C, Stiaccini G, Figliolini F, Dellepiane S, et al. Endothelial progenitor cell-derived extracellular vesicles protect from complement-mediated mesangial injury in experimental anti-Thy1.1 glomerulonephritis. Nephrol Dial Transplant. 2015;30(3):410–22. https://doi.org/10.1093/ndt/gfu364.
46. Yang J, Wang M, Zhu F, Sun J, Xu H, Chong Lee Shin OL, et al. Putative endothelial progenitor cells do not promote vascular repair but attenuate pericyte-myofibroblast transition in UUO-induced renal fibrosis. Stem Cell Res Ther. 2019;10(1):104. https://doi.org/10.1186/s13287-019-1201-5.
47. Cantaluppi V, Biancone L, Figliolini F, Beltramo S, Medica D, Deregibus MC, et al. Microvesicles derived from endothelial progenitor cells enhance neoangiogenesis of human pancreatic islets. Cell Transplant. 2012;21(6):1305–20. https://doi.org/10.3727/096368911X627534.
48. Li X, Jiang C, Zhao J. Human endothelial progenitor cells-derived exosomes accelerate cutaneous wound healing in diabetic rats by promoting endothelial function. J Diabetes Complicat. 2016;30(6):986–92. https://doi.org/10.1016/j.jdiacomp.2016.05.009.
49. Zhang J, Chen C, Hu B, Niu X, Liu X, Zhang G, et al. Exosomes derived from human endothelial progenitor cells accelerate cutaneous wound healing by promoting angiogenesis through Erk1/2 signaling. Int J Biol Sci. 2016;12(12):1472–87. https://doi.org/10.7150/ijbs.15514.
50. Xu J, Bai S, Cao Y, Liu L, Fang Y, Du J, et al. miRNA-221-3p in endothelial progenitor cell-derived exosomes accelerates skin wound healing in diabetic mice. Diabetes Metab Syndr Obes. 2020;13:1259–70. https://doi.org/10.2147/DMSO.S243549.

51. Bai S, Yin Q, Dong T, Dai F, Qin Y, Ye L, et al. Endothelial progenitor cell-derived exosomes ameliorate endothelial dysfunction in a mouse model of diabetes. Biomed Pharmacother. 2020;131:110756. https://doi.org/10.1016/j.biopha.2020.110756.
52. Wu X, Liu Z, Hu L, Gu W, Zhu L. Exosomes derived from endothelial progenitor cells ameliorate acute lung injury by transferring miR-126. Exp Cell Res. 2018;370(1):13–23. https://doi.org/10.1016/j.yexcr.2018.06.003.
53. Zhou Y, Li P, Goodwin AJ, Cook JA, Halushka PV, Chang E, et al. Exosomes from endothelial progenitor cells improve outcomes of the lipopolysaccharide-induced acute lung injury. Crit Care. 2019;23(1):44. https://doi.org/10.1186/s13054-019-2339-3.
54. Lu J, Yang J, Zheng Y, Chen X, Fang S. Extracellular vesicles from endothelial progenitor cells prevent steroid-induced osteoporosis by suppressing the ferroptotic pathway in mouse osteoblasts based on bioinformatics evidence. Sci Rep. 2019;9(1):16130. https://doi.org/10.1038/s41598-019-52513-x.
55. Jia Y, Zhu Y, Qiu S, Xu J, Chai Y. Exosomes secreted by endothelial progenitor cells accelerate bone regeneration during distraction osteogenesis by stimulating angiogenesis. Stem Cell Res Ther. 2019;10(1):12. https://doi.org/10.1186/s13287-018-1115-7.
56. Cui Y, Fu S, Sun D, Xing J, Hou T, Wu X. EPC-derived exosomes promote osteoclastogenesis through LncRNA-MALAT1. J Cell Mol Med. 2019;23(6):3843–54. https://doi.org/10.1111/jcmm.14228.
57. Zhou Y, Li P, Goodwin AJ, Cook JA, Halushka PV, Chang E, et al. Exosomes from endothelial progenitor cells improve the outcome of a murine model of sepsis. Mol Ther. 2018;26(5):1375–84. https://doi.org/10.1016/j.ymthe.2018.02.020.
58. He Z, Wang H, Yue L. Endothelial progenitor cells-secreted extracellular vesicles containing microRNA-93-5p confer protection against sepsis-induced acute kidney injury via the KDM6B/H3K27me3/TNF-α axis. Exp Cell Res. 2020;395(2):112173. https://doi.org/10.1016/j.yexcr.2020.112173.
59. Ma X, Wang J, Li J, Ma C, Chen S, Lei W, et al. Loading MiR-210 in endothelial progenitor cells derived exosomes boosts their beneficial effects on hypoxia/Reoxygeneation-injured human endothelial cells via protecting mitochondrial function. Cell Physiol Biochem. 2018;46(2):664–75. https://doi.org/10.1159/000488635.
60. Wang J, Chen S, Ma X, Cheng C, Xiao X, Chen J, et al. Effects of endothelial progenitor cell-derived microvesicles on hypoxia/reoxygenation-induced endothelial dysfunction and apoptosis. Oxidative Med Cell Longev. 2013;2013:572729. https://doi.org/10.1155/2013/572729.
61. Yue Y, Wang C, Benedict C, Huang G, Truongcao M, Roy R, et al. Interleukin-10 deficiency alters endothelial progenitor cell-derived exosome reparative effect on myocardial repair via integrin-linked kinase enrichment. Circ Res. 2020;126(3):315–29. https://doi.org/10.1161/CIRCRESAHA.119.315829.
62. Ubil E, Duan J, Pillai IC, Rosa-Garrido M, Wu Y, Bargiacchi F, et al. Mesenchymal-endothelial transition contributes to cardiac neovascularization. Nature. 2014;514(7524):585–90. https://doi.org/10.1038/nature13839.
63. Lin F, Zeng Z, Song Y, Li L, Wu Z, Zhang X, et al. YBX-1 mediated sorting of miR-133 into hypoxia/reoxygenation-induced EPC-derived exosomes to increase fibroblast angiogenesis and MEndoT. Stem Cell Res Ther. 2019;10(1):263. https://doi.org/10.1186/s13287-019-1377-8.
64. Ke X, Yang D, Liang J, Wang X, Wu S, Hu C. Human endothelial progenitor cell-derived exosomes increase proliferation and angiogenesis in cardiac fibroblasts by promoting the mesenchymal-endothelial transition and reducing high mobility group box 1 protein B1 expression. DNA Cell Biol. 2017;36(11):1018–28. https://doi.org/10.1089/dna.2017.3836.
65. Liu W, Zhang H, Mai J, Chen Z, Huang T, Wang S, et al. Distinct anti-fibrotic effects of exosomes derived from endothelial Colony-forming cells cultured under Normoxia and hypoxia. Med Sci Monit. 2018;24:6187–99. https://doi.org/10.12659/MSM.911306.
66. Izumiya Y, Araki S, Usuku H, Rokutanda T, Hanatani S, Ogawa H. Chronic C-type natriuretic peptide infusion attenuates angiotensin II-induced myocardial superoxide production and cardiac remodeling. Int J Vasc Med. 2012;2012:246058. https://doi.org/10.1155/2012/246058.

67. Valente AJ, Clark RA, Siddesha JM, Siebenlist U, Chandrasekar B. CIKS (Act1 or TRAF3IP2) mediates angiotensin-II-induced Interleukin-18 expression, and Nox2-dependent cardiomyocyte hypertrophy. J Mol Cell Cardiol. 2012;53(1):113–24. https://doi.org/10.1016/j.yjmcc.2012.04.009.
68. Gu S, Zhang W, Chen J, Ma R, Xiao X, Ma X, et al. EPC-derived microvesicles protect cardiomyocytes from Ang II-induced hypertrophy and apoptosis. PLoS One. 2014;9(1):e85396. https://doi.org/10.1371/journal.pone.0085396.
69. Kunter U, Rong S, Djuric Z, Boor P, Müller-Newen G, Yu D, et al. Transplanted mesenchymal stem cells accelerate glomerular healing in experimental glomerulonephritis. J Am Soc Nephrol. 2006;17(8):2202–12. https://doi.org/10.1681/ASN.2005080815.
70. Rampino T, Gregorini M, Bedino G, Piotti G, Gabanti E, Ibatici A, et al. Mesenchymal stromal cells improve renal injury in anti-thy 1 nephritis by modulating inflammatory cytokines and scatter factors. Clin Sci (Lond). 2011;120(1):25–36. https://doi.org/10.1042/CS20100147.
71. Lee DY, Cho TJ, Kim JA, Lee HR, Yoo WJ, Chung CY, et al. Mobilization of endothelial progenitor cells in fracture healing and distraction osteogenesis. Bone. 2008;42(5):932–41. https://doi.org/10.1016/j.bone.2008.01.007.
72. Qin Y, Zhang C. Endothelial progenitor cell-derived extracellular vesicle-meditated cell-to-cell communication regulates the proliferation and osteoblastic differentiation of bone mesenchymal stromal cells. Mol Med Rep. 2017;16(5):7018–24. https://doi.org/10.3892/mmr.2017.7403.
73. Chen G, Li P, Liu Z, Zeng R, Ma X, Chen Y, et al. Enrichment of miR-126 enhances the effects of endothelial progenitor cell-derived microvesicles on modulating MC3T3-E1 cell function via Erk1/2-Bcl-2 signalling pathway. Prion. 2019;13(1):106–15. https://doi.org/10.1080/19336896.2019.1607464.
74. Garbuzova-Davis S, Kurien C, Haller E, Eve DJ, Navarro S, Steiner G, et al. Human bone marrow endothelial progenitor cell transplantation into symptomatic ALS mice delays disease progression and increases motor neuron survival by repairing blood-spinal cord barrier. Sci Rep. 2019;9(1):5280. https://doi.org/10.1038/s41598-019-41747-4.
75. Garbuzova-Davis S, Willing AE, Ehrhart J, Wang L, Sanberg PR, Borlongan CV. Cell-free extracellular vesicles derived from human bone marrow endothelial progenitor cells as potential therapeutics for microvascular endothelium restoration in ALS. NeuroMolecular Med. 2020;22(4):503–16. https://doi.org/10.1007/s12017-020-08607-1.
76. Das CK, Jena BC, Banerjee I, Das S, Parekh A, Bhutia SK, et al. Exosome as a novel shuttle for delivery of therapeutics across biological barriers. Mol Pharm. 2019;16(1):24–40. https://doi.org/10.1021/acs.molpharmaceut.8b00901.
77. Fan H, Goodwin AJ, Chang E, Zingarelli B, Borg K, Guan S, et al. Endothelial progenitor cells and a stromal cell-derived factor-1α analogue synergistically improve survival in sepsis. Am J Respir Crit Care Med. 2014;189(12):1509–19. https://doi.org/10.1164/rccm.201312-2163OC.
78. Sun R, Huang J, Sun B. Mobilization of endothelial progenitor cells in sepsis. Inflamm Res. 2020;69(1):1–9. https://doi.org/10.1007/s00011-019-01299-9.
79. Li Y, Wang J, Chen S, Wu P, Xu S, Wang C, et al. miR-137 boosts the neuroprotective effect of endothelial progenitor cell-derived exosomes in oxyhemoglobin-treated SH-SY5Y cells partially via COX2/PGE2 pathway. Stem Cell Res Ther. 2020;11(1):330. https://doi.org/10.1186/s13287-020-01836-y.
80. Wang J, Chen S, Zhang W, Chen Y, Bihl JC. Exosomes from miRNA-126-modified endothelial progenitor cells alleviate brain injury and promote functional recovery after stroke. CNS Neurosci Ther. 2020;26(12):1255–65. https://doi.org/10.1111/cns.13455.
81. Yi M, Wu Y, Long J, Liu F, Liu Z, Zhang YH, et al. Exosomes secreted from osteocalcin-overexpressing endothelial progenitor cells promote endothelial cell angiogenesis. Am J Physiol Cell Physiol. 2019;317(5):C932–C41. https://doi.org/10.1152/ajpcell.00534.2018.

Index

P. V. Pham, S. Fagoonee (eds.), *Stem Cell-Derived Exosomes and Microvesicles*, Stem Cells in Clinical Applications, https://doi.org/10.1007/978-3-031-90698-5

The manufacturer's authorised representative in the EU is Springer Nature Customer Service Centre GmbH, Europaplatz 3, 69115 Heidelberg, Germany. If you have any concerns regarding our products, please contact ProductSafety@springernature.com

Printed and bound by CPI Group (UK) Ltd, Croydon, CR0 4YY
01/10/2026
02225982-0010